Get Started In
Urban Beekeeping

Teach Yourself®

Get Started In Urban Beekeeping

Claire Waring & Adrian Waring

First published in Great Britain in 2016 by John Murray Learning. An Hachette UK company.

British Library Cataloguing in Publication Data: a catalogue record for this title is available from the British Library.

ISBN: 978 147 3 61173 3

eBook ISBN: 978 147 3 61178 8

1

The publisher has used its best endeavours to ensure that any website addresses referred to in this book are correct and active at the time of going to press. However, the publisher and the author have no responsibility for the websites and can make no guarantee that a site will remain live or that the content will remain relevant, decent or appropriate.

The publisher has made every effort to mark as such all words which it believes to be trademarks. The publisher should also like to make it clear that the presence of a word in the book, whether marked or unmarked, in no way affects its legal status as a trademark.

Every reasonable effort has been made by the publisher to trace the copyright holders of material in this book. Any errors or omissions should be notified in writing to the publisher, who will endeavour to rectify the situation for any reprints and future editions.

Typeset by Cenveo® Publisher Services.

Printed and bound in Great Britain by CPI Group (UK) Ltd., Croydon, CR0 4YY.

John Murray Learning policy is to use papers that are natural, renewable and recyclable products and made from wood grown in sustainable forests. The logging and manufacturing processes are expected to conform to the environmental regulations of the country of origin.

Carmelite House
50 Victoria Embankment
London EC4Y 0DZ
www.hodder.co.uk

Contents

1

Introduction: bees in the urban environment

Do bees thrive in an urban environment? This may seem a superfluous question, given the subject of this book, but bees are extremely adaptable and will tolerate many things that we, as beekeepers, subject them to. One answer is another question: 'Why not?' And, indeed, why shouldn't they, as long as their needs for food, water and housing are met?

What we will consider in this book is what we, as beekeepers, can do to work with the bees to give them the best possible chance of thriving in an urban environment, while being aware of our neighbours and the various health and safety factors affecting both us and our bees.

Traditionally, we associate bees with the picture-postcard image of a cottage in the country, roses round the door and a hive or two tucked among the flowers in the riotous borders. This may have been the case many years ago, when farm workers and country folk might keep a couple of colonies to provide their family with honey and perhaps beeswax for candles or polish. However, life has changed and this idyllic situation has essentially disappeared. We are now in the age of large acreages of monoculture crops, with diminishing areas of hedgerows and meadows to provide the varied diet bees need. Towns and cities have expanded and we seem to be determined to cover the ground with concrete and tarmac.

Country beekeeping has changed and urban beekeeping is on the increase. No doubt another factor contributing to the rise of beekeeping in urban areas is the media coverage of the problems facing not only honeybees but also the whole range of pollinators. Honeybees have tended to steal the headlines, particularly with the huge colony losses experienced by commercial beekeepers in America, first reported in 2006. Checking their hives in the spring, the beekeepers found that a large number of their colonies had been decimated. The majority of the worker bees had simply disappeared, leaving only a queen and a small number of attendant workers. The problem was that the missing workers were simply that – missing! There was no sign of them having crawled out of the hive and died. They had simply left.

Theories about why this phenomenon, dubbed Colony Collapse Disorder (CCD), occurred have abounded. Some, such as new viruses, are feasible; others, such as that the bees had been translated up into heaven, less so. Many thousands of dollars have been put into research worldwide looking for a cause and attempting to find a cure, but all we know to date is that a number of different factors cause CCD.

The latest suspect is neonicotinoid pesticides, which are blamed for honey bee memory loss. Thus the bees fly out and then cannot find their way home. Much of the work being done is very technical and the results from different studies can appear to contradict one another. I am neither a chemist, a toxicologist, a biologist nor an animal behaviourist, so at this stage I would not like to give a definitive opinion, as I believe more work needs to be done. However, if there is one good aspect of CCD, it is that, from the research carried out, we now know a lot more about bees, their biology and behaviour. And that cannot be a bad thing. Maybe every cloud does have some sort of a silver lining, after all.

Remember this

Honeybees are the same whether they live in an urban or a rural location. Beekeeping is essentially the same in an urban environment as in a rural one, but the emphasis on certain aspects is different. The most important things you need to be aware of are good-tempered bees, swarm control, your neighbours, and health and safety.

There is no doubt that the environments for countryside bees and urban bees are different. So do urban bees have any advantages over their countryside cousins?

Firstly, they are likely to have access to a wider range of flowers over a longer blooming period. In the countryside where there is a pattern of monoculture, once the feast is over, the bees may well face a famine, or at least hard times as they search for forage in the sparse hedgerows. In an urban situation, bees can have access to parks and gardens. Whenever I fly into Heathrow Airport and look out of the window, I am always struck by the number and size of the parks within the city of London and this pattern is repeated in urban conurbations elsewhere.

Another benefit for urban bees is that the temperature is generally slightly higher in a built-up area compared with the open countryside, giving the bees an earlier and longer season and assistance in winter survival.

However, these are necessarily very generalized statements. Bees placed in the middle of a concrete jungle with no access to food and water will not survive. Before becoming an urban beekeeper, you need to assess the suitability of your location for your new charges. This is not just food and water but a generally suitable living environment. As I said, bees are extremely adaptable – within limits – but to be a successful urban beekeeper your situation must also be suitable.

In this book we will consider the bees and their requirements and also those necessary for you to be able to enjoy and be successful in your new hobby.

Key ideas

* Urban situations offer a honeybee colony a greater variety of forage over a longer period.
* The average temperature is warmer in built-up areas and this can affect the timing of the colony's annual cycle.
* By keeping bees, you are taking responsibility for living creatures and so you commit yourself to ensuring that their needs are met and they are kept as healthy as possible.

Focus points

* Bees are very adaptable and can live equally successfully in urban and rural situations.
* However, to do so, they need suitable and sufficient forage within flying distance of their hives.
* Honeybee populations are declining because of various factors, some understood and some unknown.

Next step

The next chapter takes a closer look at the honeybee itself and the component parts of the colony – the queen, the worker and the drone. The life cycle of individual bees and the colony as a whole will be discussed and the need for proper preparations if you want to become a beekeeper. We will take a general look at the life cycle of individual bees and the colony as a whole, and outline why you need to make proper preparations if you want to become a beekeeper.

2

All about the honeybee

What do you know about these insects that you wish to look after? The two commonest facts that everyone knows about bees are that they make honey (delicious) and they sting (not such good news). There is a lot more that you, as a beekeeper, need to know. I cannot emphasize strongly enough that the more you understand about the honey bee, its life cycle and behaviour, the more you will enjoy your beekeeping. Unfortunately, some people dive into beekeeping with only a sketchy, idealistic picture of what might be involved. This is not good news for the bees or the other human residents in their area. Of course, if you are reading this, that comment obviously does not apply to you!

The honeybee colony

Honeybees are what are known as 'social insects'. This means that they live in a group, the colony, and individuals cannot survive for any length of time in isolation. The colony itself can be regarded as an entity, sometimes referred to as a 'superorganism'. This means that the individuals within it perform different roles, which, together, maintain the colony's existence.

As the name implies, honeybees produce honey. They do not do this for our benefit, but for theirs. They visit flowers from which they suck up nectar, carrying it back to the nest in their honey crop (an extra stomach). Once there, they pass the nectar to worker bees in the hive, and these worker bees add enzymes, evaporate off water and turn it into honey. The honey is used as a carbohydrate food and stored to sustain the colony during the winter when there is no forage available.

As bees visit flowers, they also collect pollen. The pollen is also taken back to the nest, in loads on their back legs. Here it is used as a protein food and is a component of the food fed to the developing larvae. Flowers have evolved alongside bees and other pollinating insects as the collection and transportation of pollen between one bloom and another is the way that flowers set seed, thus ensuring future generations.

Bees also collect water and a sticky resin called propolis, which they use to plug gaps in the hive and cover over things that they cannot or do not wish to eject from the nest. (All these subjects will be covered in more detail in later chapters.)

The composition of the colony

A colony generally contains one queen bee, up to a few thousand male drone bees, and tens of thousands of female worker bees. The female sex comprises two castes, queen and worker.

THE QUEEN BEE

Most people know that a bee colony contains a single queen. (There are times when there is more than one, but that is not for consideration here.) The queen is an egg-laying machine, pure and simple. Once she has mated, her sole function is to produce more bees. Genetics is a complicated enough discipline to start with, but with bees it is even more so.

In humans, all eggs are fertilized and, depending on the combination of chromosomes, produce a boy or a girl. In bees, the queen has the ability to fertilize an egg before she lays it, or she can lay an unfertilized one. Both of these eggs can develop into viable adults. The fertilized eggs produce female (worker) bees and the unfertilized ones result in males (drones).

The majority of the colony consists of workers, with drones being produced only when they might be needed for mating with new virgin queens, but more of that later.

The queen is somewhat larger than the worker, but not massively so, as you would find in an ant colony. She must be able to fly because she mates on the wing and, if the colony swarms, she must be able to fly to the new nest site. Her abdomen is larger than that of a worker because it contains her two ovaries, full of developing eggs. She is the only female capable of mating and she does so within a week or so of emergence. Mating takes place at a drone congregation area – a place to which both drones and virgin queens fly. It's a bit like girls and boys meeting up at a nightclub or in the pub. The queen mates with 12 to 20 drones – the more the better!

The act of mating kills the drone as it involves the eversion of his endophallus (sexual parts). He falls backwards and away from the queen, leaving the bits behind. Presumably he has a smile on his face! The next drone removes the blockage before he mates, repeating the cycle. On rare occasions, the 'mating sign' can be seen at the end of the queen's abdomen as she returns to the hive. Presumably it is either pulled out by the workers or shrivels and falls out, because the passage must be clear when she starts laying eggs.

The queen stores the received sperm in a small spherical vessel called the spermatheca. This opens into the oviduct, a valve at the end allowing the queen to deposit sperm on a passing egg, or not, as the case may be. Generally, the queen receives and stores sufficient sperm to last her lifetime, although some believe that she may go on additional mating flights.

Key idea

Whether an egg is fertilized or unfertilized, what is important to grasp is that both types can still hatch, develop and turn into adults.

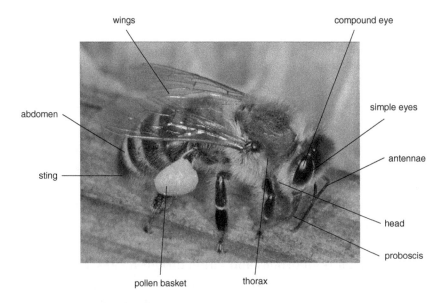

Figure 2.1 A worker western honeybee

THE WORKER

The worker bee (Fig. 2.1) is female, developing from a fertilized egg. Her abdomen contains vestigial ovaries and, in certain circumstances, she can lay eggs. However, since she cannot mate, these can produce only drones. Thus, if a colony loses its queen, fails to replace her and develops 'laying workers', it is doomed.

The workers live up to their name and are responsible for maintaining life within the colony. They maintain the brood, build comb, guard the entrance and forage for nectar and pollen.

When a worker emerges from her cell, her first task is to clean out recently vacated cells. As she matures, she moves through a series of tasks within the hive and is known as a 'house bee'. The timing of these tasks can be listed but, because we are talking about living individuals within a superorganism, this is only a rough guide. As the needs of the colony change, so workers can stay at a job longer than usual or accelerate their progression.

The usual sequence is as follows:

1 Cleaning cells (1–3 days)

2 a Brood rearing (3–15 days)

   b Looking after the queen (3–15 days)

3 a Receiving and storing nectar and pollen (3–15 days)

   b Secreting wax and building comb (10–18 days)

4 Ventilation of the nest (20+ days)

5 Guarding the nest (20+ days)

6 Foraging (20–40 days).

The worker's adult lifespan during the summer is up to 40 days, with approximately the first half being spent in the hive undertaking tasks 1 to 3. At 16–20 days old, a worker is 'middle-aged' and can switch between inside and outside tasks. This flexibility enables individuals to respond to the needs of the colony.

Over winter, a worker bee can live for around six months, but more about that later.

Worker bees are the ones that can sting you. The queen does have a sting but she reserves this for stinging rival queens.

It is smooth and can be withdrawn from the victim and reused. The worker's sting is barbed and lodges in the victim. The sting shaft has three parts: a duct from the venom sac down which the venom flows, 'zipped' together with two barbed lancets. When the shaft goes into the victim, muscles move the two lancets backwards and forwards. The barbs catch in the skin, so the 'backwards' movement of one lancet helps push the other further in. When the worker either flies off or is brushed off, the sting apparatus and intestines are pulled out of its body. The bee will die shortly afterwards.

If (when) you are stung, remove the sting as quickly as possible. Some people will tell you not to pull it out as this squeezes more venom into the wound, but research has shown that the speed of removal is more important than how it is removed.

The first couple of times you are stung, you may not see a reaction. However, with subsequent stings, the wound will itch and swell, sometimes spectacularly. For most beekeepers, their reaction gradually gets less with more stings. Be warned! You may not react much but a sting will still hurt, however long you have been keeping bees. However, some people are allergic to bee venom and it can cause anaphylactic shock. The victim has difficulty breathing and, in extreme cases, can die if adrenalin is not administered very quickly. If you prove to be allergic but wish to continue beekeeping, consult your doctor and make sure that you never go to the apiary alone. Always carry a charged mobile phone so the emergency services can be called if necessary.

THE DRONE
The third type of bee in the colony is the male drone. Drones have been described as lazy and, indeed, they do not take an active part in the colony's maintenance. Generally, they are fed by workers, although they will take nectar or honey from well-filled cells.

The drone's main function is to be available to mate with a virgin queen, thus ensuring the continuity of the species. Drones may also perform a secondary function as 'draught excluders' as they are often found congregated at the outer edges of the brood nest. Drones are responsible for the morale of the colony.

The life cycle of the honeybee

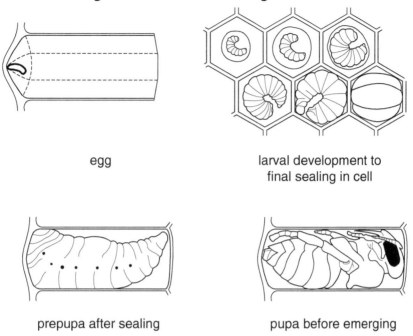

egg

larval development to
final sealing in cell

prepupa after sealing

pupa before emerging

Figure 2.2 The life cycle of the honeybee

Every bee begins life as an egg laid by the queen (Fig. 2.2). Eggs are laid in the hexagonal cells of the beeswax comb. Cells are built on either side of a beeswax midrib and are offset so that the bases interlock. In a natural situation, such as a swarm occupying a natural cavity, comb is built in a continuous sheet from top to bottom. Because beekeepers want to manipulate a colony, they use wooden frames containing sheets of beeswax embossed with the cell pattern (foundation). This gives the bees a start and also helps ensure that they build comb where we want them to. Worker cells are the most numerous and are smaller than those used to raise drones. Queen cells are different, hanging down from the face of the comb.

All eggs hatch into larvae after three days, but then the development of the different bees starts to diverge. There are two types of brood food – worker jelly and royal jelly. Initially, all larvae receive royal jelly. After the second day, worker larvae move on to worker jelly but queen larvae continue to receive royal jelly. Queen larvae are given a lot more food than worker larvae and the higher sugar content appears to encourage them to eat more as well, thus making them grow more quickly.

The larvae then metamorphose into adults. At this stage, the house bees seal the cells with beeswax. The cappings of worker cells are more or less flat while those of drones are domed to accommodate the larger insect. The cells of queen larvae are sealed on day 8, those of workers on day 9 and for drones it is day 10. Change into the adult form takes 8, 12 and 14 days respectively. The total developmental times are:

▶ queen – 16 days

▶ worker – 21 days

▶ drone – 24 days.

Learn these timings, especially those for the queen and worker, as they are pivotal in understanding the different methods of swarm control. If you understand the first principles, you will be in a position to take appropriate action if you meet an unexpected situation when you look at your bees.

Remember this

The development times for the queen, workers and drones are important as they impinge on other beekeeping manipulations.

The life cycle of the colony

In theory, the honeybee colony is everlasting. Individuals come and go but, as long as there is a properly mated, laying queen, the colony will continue. Of course, in practice, it is not like this, since survival relies on other factors such as food availability and the absence of disease.

As the colony year is cyclical, it is difficult to know where to begin but spring seems as good a place as any. As the days lengthen, the queen increases her egg-laying rate. This means there are more larvae to feed and the nurse bees access pollen and honey from the stores to provide this. Foragers leave the hive when the weather is suitable and collect more supplies from the early flowers.

Beekeeper or not, you can help bees at this critical time of year by planting spring-flowering bulbs such as crocus and grape hyacinth and shrubs such as male pussy willow. For those with limited garden space, the weeping Kilmarnock willow is ideal.

The colony continues to expand and surplus nectar is stored above the brood nest as honey. This is when you need to provide more storage space. Failure to do so may encourage the colony to swarm.

Swarming is a natural process, which can be thought of as colony reproduction. The queen leaves the hive with about half its bees and establishes a new colony in a suitable cavity. Research by Thomas Seeley at Cornell University, New York, has shown that bees prefer one with a capacity of approximately 40 litres (11 US gallons), 3–4 m (10–13 ft) above the ground, with a small entrance facing south. Well, we can't always have our ideal home, so scout bees investigate several cavities and estimate the volume, looking for the one that comes closest. The manmade beehive closest to the ideal 'des res' is the Langstroth design, but bees seem happy to live in other hive types, especially when the space is expanded with extra boxes.

In the old hive, new queens are raised, each leaving the hive with more bees to set up their own colonies. Eventually, the bees in the original colony 'decide' that losing more bees would jeopardize their survival and so one queen is selected and any others killed. This virgin queen mates on the wing and, on returning, heads the colony, ensuring its continuation.

The colony continues to gather and store food in readiness for the winter. Later in the summer, as day length shortens, the queen reduces her egg laying and the colony size diminishes.

As winter approaches, the bees begin to cluster together, maintaining a constant temperature at the centre. The queen lays either very few or no eggs and the bees gradually eat their way through their food stores. The honey consumed is converted to heat by the vibration of their flight muscles, which can be disconnected from the wings themselves. Bees on the outside of the cluster cool down. They then move into the middle to be replaced by 'warm' bees.

Eventually, days begin to lengthen and the cycle begins all over again.

Key idea

The colony has an annual pattern, expanding in the spring, collecting and storing honey, contracting in the autumn and living off its honey stores during the winter.

Honeybee subspecies

The honeybees kept in the UK and America are western honeybees (*Apis mellifera*). They nest in cavities to protect them from predators and help them keep warm. Other bees, notably the Asian honey bees *Apis dorsata*, *Apis laboriosa*, *Apis florea* and *Apis andreniformis*, build a single comb out in the open and utilize a curtain of bees to perform these functions.

There are four major subspecies of *Apis mellifera*:

▶ *Apis mellifera carnica* – the carniolan bee found from the Alps to Slovenia, Croatia and Serbia.

▶ *Apis mellifera caucasica* – the Caucasian bee, found in the mountainous area between the Black and Caspian seas, from southern Russia to Azerbaijan.

▶ *Apis mellifera ligustica* – the Italian bee, originating in that country but now found all around the world.

▶ *Apis mellifera mellifera* – the dark European bee of northern Europe and western Russia. This is the bee native to the British Isles, although it has been widely hybridized with imports of *Apis mellifera ligustica* and *Apis mellifera carnica*.

Other bee species

It is worth mentioning other bee species you will see in your garden. All are important pollinators but they have different lifecycles from the honeybee, and they are usually either bumblebees or solitary bees.

- **Bumblebees:** There are currently 24 species of bumblebee in the UK, the commonest being *Bombus terrestris*, the buff-tailed bumblebee. A mated queen hibernates during the winter, emerging in spring to search for a suitable nest site. Initially, she does all the work – collecting pollen and nectar, building wax cells, laying eggs, incubating them, feeding the larvae and capping their cells. As new workers emerge, they join in foraging activities and eventually the queen remains in the nest just laying eggs. Towards the end of the season (which varies according to species), new queens and drones are produced and mate. The new queens hibernate and the old nest perishes.

- **Solitary bees:** Solitary bees are what they say on the tin. Males and females emerge in the spring and mate. The females then find a suitable nest cavity, provision it with pollen, moistened with nectar, and lay an egg. The cell is then sealed and left to develop on its own. The adults emerge the following spring. Solitary bees often look for tubes such as hollow plant stems. Cells are constructed in sequence from the end to the opening.

If you want to encourage pollinators, you can buy bumblebee nest boxes and also solitary bee nests consisting of a bundle of suitable tubes. I would particularly recommend the latter, which can give you a great deal of enjoyment. Bumblebees often take over old mouse or birds' nests and I have frequently been called to a 'swarm' of bumblebees in a bird box.

Neither male bumblebees nor male solitary bees have a sting. The females do, but rarely use them. Those of solitary bees are very small and generally cannot pierce human skin. Bumblebees are mostly non-aggressive, preferring to warn you by lifting their middle leg into the air. If you persist and move closer, they will raise two legs and turn on their backs to show you their sting. However, they rarely use it. If they do, they will survive because it is not barbed. Recently, the tree bumblebee (*Bombus hypnorum*) has spread into the UK from the Continent. It tends to be more defensive of its nest than other species. All bees, including bumblebees, will ignore you when they are out foraging.

Bees as livestock

Long gone are the days when a colony of bees could be left to its own devices at the bottom of the garden until it was time to take off the honey crop. Our bees are now afflicted with the varroa mite (*Varroa destructor*), which feeds on the blood of the developing larva and transmits and activates deadly viruses in the process. This is in addition to the 'standard' set of potential diseases such as American foul brood, European foul brood, Nosema, Acarine and the chalkbrood fungus.

Waiting in the wings, so to speak, are other nasties. The Asian hornet (*Vespa velutina*), which can destroy a honeybee colony in a short space of time, has spread through France and is waiting for an opportunity to cross the Channel. Likewise, the small hive beetle (*Aethina tumida*), which makes a mess in the hive and can totally spoil the stored honey, has now been positively identified in colonies in southern Italy.

I suspect it is only a matter of time before these exotic pests reach the UK. However, beekeepers have proved as resilient and adaptable as the honeybee itself and I have no doubt that beekeeping will continue. It will be different, but beekeepers have overcome other challenges, such as when the acreage of oilseed rape expanded across the country and – even more so – when the varroa mite arrived.

If any of the above puts you off the idea of becoming a beekeeper, I suggest that you don't keep bees. You might have thought I would be all in favour of as many people keeping bees as possible. In one way I am, but I want those beekeepers to be well educated in the ways of their bees and competent to look after them properly 'in sickness and in health'. To keep bees well, you need to realize that they are like cattle or sheep or chickens. They are livestock, even though they fly out of the hive to do their own thing and don't need your attention several times a day. They may be able largely to look after themselves but, if you want to call yourself a beekeeper, it is your responsibility to give them the care and attention they require.

Remember this

By becoming a beekeeper, you are taking responsibility for living beings that need to be cared for.

Key ideas

* There are three types of bee in a colony: the queen, the only one capable of laying fertilized eggs; many female workers, which perform duties to maintain the colony's existence; and drones, whose primary job is to mate with virgin queens.
* Individual honeybees cannot survive for long on their own.
* Each individual follows a lifecycle pattern from egg to larva to pupa to adult. It then performs the functions for which it is suited, and eventually dies.

Focus points

* Honeybees are social insects, living together in a colony that can be regarded as a superorganism.
* The queen is essential to the colony's existence as she is the mother of all the constituent bees.
* The workers are female but are unable to mate and lay fertilized eggs. If they do lay eggs, these develop into drones.
* The drones are male and are raised as the colony expands in spring so that they are available to mate with any virgin queens that are produced.
* The honeybee colony is designed to survive from year to year. Bumblebees and solitary bees have annual lifecycles, with new queens being produced each year to carry on the species in the following one.

Next step

The next chapter will consider where you can keep your bees. It will look at the requirements for an apiary site at home, on an allotment or a roof, and how important it is to consider your neighbours. Aspects to be considered include ease of access, forage availability in the area, protection from the wind, working space and the provision of water.

3

Siting an apiary

An apiary is defined as 'a place where bees are kept; a collection of beehives'. This means anywhere you keep your bees. You may have space in your back garden or decide to keep hives on the flat roof of your garage. Many hotels and public buildings now have rooftop apiaries. You will also find apiaries on allotments, where local regulations allow, or sited on areas of wasteland. In one sense anything goes, but in an urban situation you have to take several other factors into account when siting an apiary. You need to think about the amount of space you have, ease of access, shelter from the wind, and the amount of forage and water that will be available for the bees. Above all, you need to consider your neighbours: trouble with bees and neighbours is something that cannot be mended.

The urban garden

Urban gardens are usually small, which means that your neighbours are very close. If you plan to site an apiary in your garden, you must think about the impact your bees may have on others. When you open the hive and look at your bees, you will be wearing protective clothing, including a veil over your face. Modern veils are excellent and, absorbed in the fascinating task in hand, you may be totally unaware of how your bees are reacting. They may not be stinging you, but they may have gone next door and be attacking your neighbours.

Your neighbours have a right to enjoy their property and will not be pleased if they have to stay indoors on a sunny day because you are inspecting your bees. Although you can take steps to minimize any disturbance to the neighbourhood, a tiny garden is not the best place for beehives. Some urban councils, especially in the USA, do not allow beekeeping within the city limits, although this rule is gradually being relaxed in many places in the light of declining bee populations.

Neighbours can be bothered by swarms that leave your hives and settle in their garden. Avoid this situation by understanding, learning and practising a method of swarm control. Inspect your colonies regularly during the swarming season and institute control measures as soon as swarming preparations begin. Different colonies also have different characteristics (just like humans) and some are more prone to swarming than others.

Some bees are very defensive of their nest (or aggressive from the human point of view), while others are far less so. A way to avoid problems with your neighbours and make your own beekeeping much more pleasurable is to keep good-tempered, low-swarming bees. I would also recommend that you obtain local bees that are adapted to your environment, but more of that later.

An allotment

An allotment or a piece of waste ground can be good, but first check that beekeeping is allowed. If it is allowed on your allotment, your fellow allotment holders will benefit from enhanced crop pollination. However, the same considerations apply as for gardens: the allotments may occupy a large open space but your next-door plot holder is still your neighbour.

One problem with keeping bees away from your home is the possibility of vandalism or theft. Some people seem to delight in destroying other people's property and when there is risk of danger, like being stung, they seem even more intent on doing so. I have heard of hives being overturned, shattered by stones thrown at them or even having lighted fireworks pushed inside. I find it very sad when I hear of beehives being stolen because it almost invariably means that this is the work of a fellow beekeeper. To close up and secure a beehive for transportation means you have to know what you are doing, and so it is unlikely to be anyone without the necessary skills.

Again, you have to assess the risks of putting your hives on an allotment. The risk is even higher on waste ground. A securely fenced area is best. At the very first hint of trouble, you are strongly recommended to move your bees elsewhere as soon as possible. A pot of honey can help sweeten the situation, but trouble between bees and your neighbours is to be avoided at all costs.

Remember this

Check with the authorities before siting hives on an allotment. Not all of them allow this. If you do decide to site your hives on your allotment, inform the other tenants and make sure that they are happy for you to do so.

A bee shed

Another alternative is to keep your bees in a shed. This keeps them out of sight of your neighbours and potential vandals. You can also inspect them easily, even if it is raining. You can easily adapt a garden shed for the purpose. You will need a strong bench for the hives at a suitable height. Make entrance holes in the wall and butt the hives up to them or make a tunnel for the bees to use so that they don't miss the hive and start flying round inside the shed.

During your inspection, bees will start flying. They will fly to the light so you need windows in your shed, preferably on the same side as the hives so you can see what you are doing. Adapt the windows to allow bees in the shed to escape and return to their hive through the entrance. A bee on a window flaps its wings and tries to climb upwards. However, it invariably slips back down and has to start again. By providing a gap at the bottom of the window, it will eventually fall out. Probably the easiest way is to create a secondary frame, which holds the glass away from the shed wall by about a centimetre. Marking the different hive entrances in different colours or with different geometric shapes helps returning bees find their own hive.

Remember this

Bees must be sited so that they give minimum disturbance to your neighbours. Ideally, they should not realize you are a beekeeper unless you tell them. You will need extra space during manipulations such as swarm control, so factor this into your apiary site from the beginning.

Hive stands

Your hives need to be set on strong, stable stands. These should be of a height that brings the top of the brood box to the level of your hands. Many beekeepers complain of 'beekeeper's back', which is caused by bending over a hive for long periods as well as by not lifting heavy boxes correctly. In the height of summer, with a good honey crop, a hive can weigh over 100 kg (220 lbs). If your fattest friend can stand on it, your stand should cope with the hive!

Hive stands come in many designs. Some fold up for easy transportation and storage when not in use. One of the simplest is made by supporting two square-section rails on concrete blocks or by fixing a square-section frame on sturdy legs. I use stands made from square-section metal, welded together. If you want to put two hives on the same stand, you need rails about 1.5 m (5 ft) long because you must have a gap between the hives.

Your hive stands need to be level both from front to back and side to side. This will encourage the bees to build comb vertically in the frames. If the hive is tilted, the comb will be vertical but this will not coincide with the frame, making manipulations difficult.

Rain should not blow into the entrance of a level hive in a sheltered apiary except in exceptional circumstances. If you wish, you can place a 5-mm ($\frac{3}{16}$-in.) block under the back edge of the hive to ensure that any water runs out of the entrance.

Remember this

Position the hive stands in your apiary so that you have adequate working room and can manipulate the colonies easily.

Other siting factors

Having taken all the above factors into account, what else do you need for your ideal apiary? Considering the following features carefully will benefit both you and your bees.

SPACE

Consider whether you have sufficient space for your apiary. Even if you want to keep only one or two hives, you will still need working room and enough space to put hive parts down while you inspect the brood box. Almost all methods of swarm control require the colony to be split into two (or more) and during this manipulation you must have room to accommodate additional hives. As a rule of thumb, I would recommend an area for each hive that is nine times its footprint. This is just a guide, though, since you can reduce the amount of space required by keeping two hives side by side on one stand.

You need room to stand by your hive so that the frames are parallel to your body. Reaching across a hive to grasp the far end of the frame is really awkward and it means that you are continually moving your hand over the bees and alerting them to your presence. You also need space nearby to put down the roof and any additional honey super boxes. If you have only two colonies, you can place them side by side on a stand with the entrances facing the same way. However, with more hives you need to vary the direction they face, and there are many different ways of doing this.

Foraging bees learn where their hive entrance is in relation to its surroundings. You can help by marking each hive with a distinctive symbol or colour pattern. Placing objects among the hives also helps the bees to orientate correctly.

ACCESS

The next very important factor is ease of access. You yourself may be fine climbing a ladder or clambering out of a bedroom window to get on to your garage roof. That's probably OK when it is just you and maybe your smoker and hive tool. However, it is a totally different matter when you want to take a heavy, awkward box of frames containing honey back into the house to extract it. You must make absolutely sure that you have safe access to your bees. You also need to make sure that others have easy access if they need to come to help you in an emergency.

Keeping bees on a rooftop is fine (Fig. 3.1), although you need to consider whether there will be enough shelter from the wind. More importantly, you must ensure that you can't fall off. Accidents happen. You may trip or slip. Your bees may get upset and start stinging you. If you panic, will you be safe?

I know that we sometimes feel that health and safety has gone over the top, but you are responsible for assessing the risk factors associated with your choice of apiary.

Key ideas

* Apiaries can be located in various locations, but it is your responsibility to determine that these are suitable, taking particular care that your bees do not adversely affect your neighbours.
* It is probably impossible to find the ideal apiary site. When looking for yours, list the requirements in priority order and fulfil as many as you can.

SHELTER FROM THE WIND

One of the most important things, especially on a rooftop, is to make sure that your bees have shelter from the wind. In a very high wind the hive itself might be blown over the edge, so you must secure it firmly in a way that also allows you to inspect it easily. Straps can be secured to the roof and placed over the hive. These will have to be adjustable for when you have extra honey super boxes in place.

Figure 3.1 Langstroth beehives in a rooftop apiary, Brooklyn, New York

Bees can only fly under control when they are going faster than the surrounding air and are only able to land easily when facing into the wind. They cannot fly in winds of more than 19–24 km (12–15 miles) per hour. If a bee is knocked down by a cold wind, it will not be able to recover and get back into the colony. If its temperature falls to 9–10 °C (48–50 °F) it cannot move sufficiently to warm up and it will die.

In a garden, a thick hedge makes a better windbreak than a solid fence. The hedge slows down the wind speed at ground level for a distance of 40 times the hedge height. A solid fence creates eddies which can blow the bees all over the place.

A barrier is also useful in terms of potential problems with your neighbours. If bees coming out of the hive are made to fly upwards to clear the barrier, they will continue flying at this height unless brought lower by the wind. They will thus fly over the heads of people nearby and are much less likely to cause a nuisance. Siting hives within a shrubbery works well as it makes the bees fly up and also gives them a signpost to their own hive when they return.

On a roof, it is not practical to provide a hedge but you can site your hives in a sheltered corner or erect a permeable barrier like a hurdle. Make sure that this is also secure!

Colonies can benefit if their hives are exposed to the sun in autumn, winter and spring. The small amount of warmth can enable bees to leave the winter cluster, fly out from the hive and defecate. However, in the height of summer (yes, even in the UK there can get heat waves!), hives can overheat, especially on a hot roof. Bees can ventilate and cool the hive to a certain extent, but this is a waste of their energy. The ideal situation in a garden is to place hives in the shade of deciduous trees, which lose their leaves in winter. In an exposed position like a roof, you may have to consider providing some sort of temporary shading. Depending on the roofing material, the surface may become very hot, so set your hives on some sort of insulation to prevent heat radiating upwards.

If bees could choose, their ideal nest would have a south-facing entrance. This may not be practical in your situation and, if your apiary is well sheltered from the wind, it will not matter too much, so you can position your hives to suit yourself. However, the positioning of your hives can be important. A regular pattern, such as placing all the hives in a straight line, confuses bees and returning foragers may try to enter or drift into the wrong hive. This is one of the main ways in which diseases are spread.

AVAILABLE FORAGE

Before choosing your apiary site, have a look around the area. To thrive and give you a good crop of honey, your bees must have access to a large amount of nectar-producing flowers. Ideally, these should bloom over an extended period, providing early nectar and pollen for the expanding brood nest in the spring, a large volume of blossoms in the summer (for the honey crop) and later sources to top up the winter stores.

Urban beekeepers can have an advantage over rural beekeepers. Although urban gardens may be small, there are many of them in a concentrated space, giving a large potential forage source. Parks are also excellent foraging grounds. Imagine a mature sycamore or lime tree spread out on the ground. It would cover a very large area. This is the 'area' of forage available to your bees in a vertical plane. In the country, bees may have access to vast acreages of oilseed rape or field beans, but when these crops have finished flowering there can be a complete dearth.

Check whether there are already other beekeepers in the area. Remember, their bees will be foraging in the same area and competing for the food. If there are too many bee colonies in an urban area, they cannot all be supported by the available forage. If you want your bees to do well, don't place them close to other apiaries.

WATER

Bees are living creatures and, as such, need water. They transport it back to the hive in their honey crops and use it for diluting honey stores and cooling the colony. It is not stored in the combs, so bees must have access to a constant water source. You can provide this with a water fountain, which can take several forms. Your water fountain should be a little way from the hives as bees naturally seek water at a distance from the nest. A tap in your apiary can be left dripping very slowly into a container of soil or peat. Put holes in the container so that the contents do not become waterlogged. Bees will drink directly from the tap or will suck up water from the wet soil.

You can provide a shallow container full of water. Place it in the sun to warm the water. If a bee drinks cold water, it will be chilled and its temperature may drop sufficiently for it not to be able to warm up and it will die. The other danger is drowning, so place stones that break the surface or wooden floats in the container, so that the bees can land safely. Above all, do not let your water fountain dry up. This is particularly important in urban areas because, if they cannot get water from the source you provide, the bees will look elsewhere and start drinking from your neighbour's pond, bird bath or other water source, causing a nuisance.

How do you train your bees to use your water fountain? The trick is to start by filling it with dilute sugar syrup (equal parts of sugar and water). Trickle a little over the top bars of the frames in the hive to encourage scout bees to fly out and hunt for the source – your water fountain. When they have located it, top it up with plain water. As you continue to do this, you can gradually reduce the sugar content but the bees will continue to drink there.

Remember this

Provide a constant source of fresh water near the apiary. Once bees start using it, it is very important that it does not dry out. If that happens, the bees will look for water elsewhere, which could be at your neighbour's pond.

Try it now

* Look for a suitable apiary site before obtaining your bees, so that you can plan properly for the placement of the hives and access.
* Obtain or construct sturdy, stable hive stands that will raise the hive to around the level of your hands.
* Make sure that there is safe access to the apiary.
* Be very conscious of the neighbours in an urban situation. They have a right to enjoy their property.
* In a built-up area with small gardens, consider keeping your bees in a bee shed.

Focus points

* Apiary sites must be chosen carefully with various factors in mind.
* In an urban situation, the prime consideration must be to your neighbours, making sure as far as is humanly possible that your bees do not cause them distress or harm.
* You must have sufficient space not only for the current number of hives but also for the additional hives you will require during swarm control manipulations.
* You need to consider the bees' requirements for adequate forage in the area and for a constant source of water.
* Hive stands need to be strong enough to support the heaviest hive that can be anticipated.

Next step

The next chapter looks at the options available for providing a suitable home for your bees, including the importance of the bee space. We will describe the different parts of the hive and give details about the construction of frames and hive boxes. Information is also given on how to clean and sterilize hives and on the care needed when acquiring second-hand equipment.

4

Hives and their components

When humans first discovered that bees make and store sweet honey, they would simply rob any nest they found, either driving the bees out or even killing them. However, that meant travelling into the forest to find them. Soon they realized that there was a better way – by providing suitable cavities near to their settlements. This exploited the fact that the honeybee (*Apis mellifera*) naturally establishes its nest in a cavity, where it builds parallel beeswax combs in which to rear brood and store pollen and honey. Either a swarm would take up residence in a cavity spontaneously or the beekeeper would bring a colony back from the wild and install it, providing easy access to a honey supply.

Honey hunters and early hives

So-called primitive hives were made from hollow logs or cylinders made from locally available plant materials. In Egypt they made hives from clay, which served the same purpose. In Nepal cavities were left in the thick house walls and blocked up on the inside and outside (a bit like bricking up a window but leaving the space). An entrance hole in the outside enabled the bees to access the cavity, while a removable plate at the back meant that the beekeeper could access the nest to harvest the honey.

The bees would then build natural comb in the container or cavity and at harvest time the beekeeper simply removed these. Some combs would contain brood as well as honey but this provided a source of protein. The bees may have remained *in situ* and built more comb, but often the disturbance caused by the beekeeper taking the honey would make them fly off to look for another nest site, meaning the beekeeper had to start all over again. Beekeepers began to realize that, if they only took some of the combs (primarily the honeycombs), the bees were much more likely to stay put.

There were several attempts to persuade bees to build their combs on removable wooden bars placed across the cavity's diameter. These were partially successful, but free-hanging comb is fragile and can easily break off.

Full-frame and top-bar hives

The next development was a full frame with top, side and bottom bars. Initially, these were made to fit the box with only a small gap between the side bars and the walls. The bees did what comes naturally – they filled the gap with propolis, a sticky resin that they collect from tree buds. This glued the frames to the hive, which rather defeated the purpose!

Observation had shown that bees naturally leave a space of 6–8 mm (¼–⁵/₁₆ in.) between their parallel combs in the honey storage area. This gap enables individual bees to store and process nectar and access the honey when required. However, bees also leave this gap between the edge of the comb and the inside walls of the cavity. In the brood nest area, and between the bottom edge of the combs and the floor, this space is doubled. This means that worker bees can pass one another when tending to the brood. Greek beekeepers using top bars also observed that, if the walls sloped slightly inwards, bees did not attach their combs to them. This knowledge has been incorporated into hives such as the Kenyan top-bar hive, widely used in Africa. Similar top-bar hives are increasing in popularity in the UK and America.

Key idea

Beekeepers provide bees with hives containing movable frames. This is purely for the beekeeper's benefit. The bees are happy with a simple cavity.

The bee space and the modern hive

The revolution came when an American clergyman, the Reverend L. L. Langstroth (1810–95), realized that a space had to be left between the frame side bars and the box, and the concept of the 'bee space' was born (Fig. 4.1). All modern frame or top-bar beekeeping is based on Langstroth's discovery. The only essential difference between different designs of frame hives is the volume of the boxes.

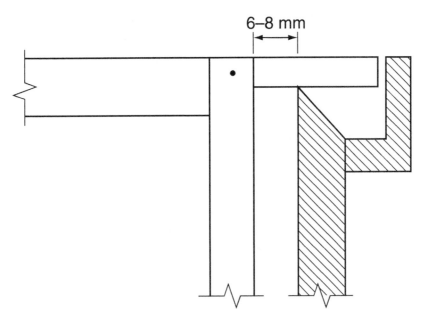

6–8 mm

Figure 4.1 The bee space between a frame and the hive

A modern hive consists of a floor, a brood box, a queen excluder, super(s), an inner cover or crownboard, and a roof.

THE FLOOR
The floor is placed on a hive stand at a comfortable working height. Previously, floors were made of solid wood but, with the advent of the varroa mite, there has been a move towards using open-mesh floors. These consist of a frame round a mesh panel with a removable tray that slides underneath. This allows monitoring of natural mite mortality as dead mites and those knocked off the bees fall through the mesh and collect on the tray. Living mites cannot climb back into the hive so the mesh floor also helps to reduce the mite population. (Pests and diseases are covered in more detail elsewhere.)

Fillets on three sides of the floor leave an entrance the width of the fourth side, formed when the brood box is placed on the floor. In most commercially available floors, the fillets are 22 mm (⁷⁄₈ in.) deep but I recommend you use a shallow floor roughly a bee space high (6–8 mm or ¼–³⁄₈ in.). With a shallow floor, bees are less inclined to build comb below the bottom bars of the brood frames. The other advantage is that mice cannot squeeze through this small gap when they are looking for somewhere warm to spend the winter. If you use a deep floor, you will have to prevent this by attaching a mouseguard over the entrance. This has a series of 10-mm (³⁄₈-in.) holes drilled along its length that allow the bees to fly in and out but keep out the rodents. An entrance block reduces the width of the deep floor. It fills the gap and has a depression, usually in the centre, measuring approximately 10 mm (³⁄₈ in.) high and 75–100 mm (2 ¹⁵⁄₁₆ – 3 ¹⁵⁄₁₆ in.) wide. Make sure that it doesn't fit too tightly, as you may have to remove it in the summer if the weather turns hot. A colony should be able to defend the smaller entrance.

THE BROOD BOX

Working upwards, the next component is the brood box (Fig. 4.2). This holds the frames and combs where the queen lays her eggs. The combs are also used to store pollen and honey, which the nurse bees use to feed the developing larvae. The frames are supported on a pair of metal runners fixed on opposite sides. The curved lip along the length fits over the hive wall.

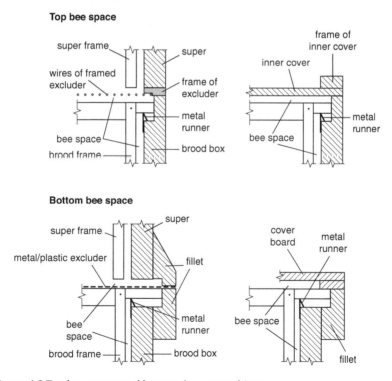

Figure 4.2 Top bee-space and bottom bee-space hives

Hives are designed to have either a top bee space or a bottom bee space between boxes. In a top-bee-space hive, the frames are situated so that there is a bee space between the top bars and the top of the box. The bottom bars are flush with the bottom of the box. The opposite is true for bottom-bee-space hives, where the top bars of the frames are flush with the top of the box and the hive walls and the bee space is situated between the bottom bars and the bottom of the box.

Different hives use different spacings, of course. Langstroth, Smith and Dadant are top bee space, and the WBC, National and Commercial are bottom bee space. There can be heated debates as to which is better. Personally, I don't think it matters that much: in both cases you are leaving a bee space gap between the frames, which is what bees need.

What you must definitely *not* do is mix two hive types. If you put a top-bee-space box on top of a bottom-bee-space box, the bottom bars in the top box will touch the top bars in the lower one and the bees will glue them together with propolis. In the reverse situation, you will create a double bee space between the boxes in which the bees will build brace comb. Both situations are bad news when you try to part the boxes to inspect the bees.

The important measurements in a hive are the internal ones. The bees don't care what the hive looks like from the outside, but they do care that the bee space is correct inside. Of course, having uniform external dimensions means it is easier to match up the internal cavities. The other important thing is to always make sure that the hive is 'bee-tight' – that is, that the only way into the hive is through the entrance. There must be no other gaps anywhere through which wasps and robber bees can squeeze – and they can get through what look like impossibly small spaces. Remember, the bee space can be as small as 6 mm (¼ in.).

THE QUEEN EXCLUDER
The queen excluder (Fig. 4.3) goes on top of the brood box. This is a framed grid or a sheet of slotted metal or plastic of the same dimensions as the hive cross section. Worker

bees can pass through the gaps but these are too small to allow the passage of the queen or drones. Thus the queen is confined to the brood box and you don't get her laying eggs in the honey supers.

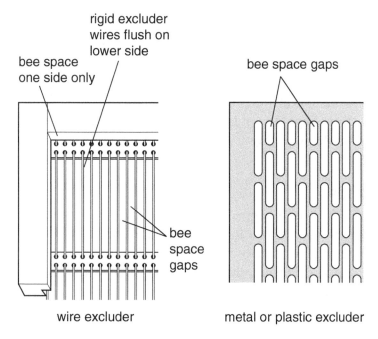

Figure 4.3 Types of queen excluder

HONEY SUPERS

The bees' natural inclination is to store honey above the brood nest. A wild colony in a natural cavity achieves this by extending the comb downwards. Bees cannot do this in a hive, so the beekeeper must give additional space by stacking super boxes on top of the queen excluder as required, extending the nest upwards. Supers are generally shallower than the brood box and take frames to fit their dimensions. You should aim to have three or four supers for each of your colonies. Nectar contains roughly 60 per cent water, which is reduced to around 18 per cent when it is converted to honey. The bees therefore need more space to store nectar than honey. They will use the super to unload nectar into the cells so that they can go out and collect some more. House bees will then work to convert the nectar to honey and eventually fill up all the super combs (in a good year!).

THE INNER COVER

The inner cover, or crownboard, is placed over the top super, or the brood box when there are no supers on the hive. It is a flat board with the same cross-sectional dimensions as the boxes. There are fillets round all four edges on both sides and usually one or two holes in the centre. These are used when you want to feed the colony. The inner cover doubles up as a 'clearer board', which is used to remove bees from the supers when you want to extract the honey. A device called a Porter bee escape fits into the hole(s). This one-way gate allows bees to push through a couple of springs on the way down but prevents them getting back through the small gap between the ends.

THE ROOF

Finally, the hive is topped with the roof. It has ventilation holes in the centre of each side, which allow air to flow through the hive, helping to keep the hive dry and prevent mould growth. Inside, the holes are covered with a bee-proof mesh to prevent unwanted visitors gaining access. The roof is protected with a cover made from metal or other suitable waterproof material.

Choosing your hive

Your choice of hive is personal. If you are just starting out, I suggest you talk to other beekeepers in your area (join your local beekeeping association to meet them) to find which is their most popular hive. This gives you a good idea of the size of hive that suits your local bees and you will also find second-hand equipment more readily available. If your association has an apiary with different hive types, you can 'try before you buy'.

THE LANGSTROTH

The Langstroth is most popular frame hive worldwide (see Fig. 3.1). It has a deep brood box measuring 508 × 413 × 243 mm (20 × 16 ¼ × 9 ⁹⁄₁₆ in.), a medium box 170 mm (6 ¹¹⁄₁₆ in.) deep and a super at 150 mm (5 ¹⁵⁄₁₆ in.). Some beekeepers prefer to use the same-sized box throughout, so that all frames are interchangeable. The Langstroth hive is a single-walled hive, constructed from four pieces of wood, jointed at the corners. Rebates in the top of the walls of the two shorter sides support the frames. Finger grips are cut out of the thickness of the timber, so lifting a heavy box can be hard on the fingertips.

THE WBC

In the UK, the style most people recognize as a 'typical' beehive is the WBC, designed by William Broughton Carr in 1890. This is a double-walled hive with the boxes containing the frames surrounded by 'lifts', which stack on top of each other and support the roof. This design may be more suitable for colder areas, giving a degree of additional protection against the elements. It is more cumbersome to work as the lifts have to be removed first. It is also difficult to move when it contains bees as the inner boxes are usually made of less substantial timber and the fit does not need to be as precise as for a single-walled hive. However, it is decorative and would look good in an urban garden. Indeed, some gardeners purchase just the outer components and use them as storage containers!

THE MODIFIED NATIONAL

Most beekeepers in the UK use the single-walled Modified National hive (commonly referred to as the National; Fig. 4.4). This is smaller than the Langstroth, measuring 460 mm (18 1/8 in.) square and 225 mm (8 7/8 in.) deep. Supers are 150 mm (5 15/16 in.) deep. For those wanting a larger brood box, the 14" × 12" is also available. The original National hive was constructed in the same way as the Langstroth. However, this design was subsequently modified, with fillets being introduced to the outside of the top and bottom of two opposite sides whose height is less than that of the other two walls. The frames are suspended from runners attached to the shallower walls. The advantage of the National hive is that it is square and the frames can therefore be orientated either parallel to (warm way) or at right angles (cold way) to the entrance (Fig. 4.4(a) and (b)).

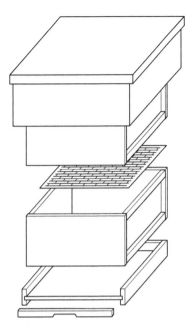

(a) Modified National hive, used the warm way. Expanded to show the floor, entrance block, brood chamber, queen excluder, super and roof

(b) Modified National hive, used the cold way

Figure 4.4 The Modified National hive

OTHER HIVES

Other commercially available hives include the Commercial, the Smith and the Dadant. Commercial brood boxes are larger than those of the National hive but you can put National supers over a Commercial brood box. The Smith hive takes the same size frames as the National, but the frames have a shorter lug. This means that, if you want to transfer frames from a National to a Smith hive, you will have to do some carpentry work. A strong pair of secateurs is usually effective. The Dadant hive is much larger than the National and is uncommon in the UK.

The Dartington Long Hive essentially consists of brood boxes combined horizontally rather than vertically. It takes up to 17 14" × 12" frames. The super boxes take six frames and are relatively easy to handle. The Beehaus is a plastic hive based on the Dartington Long Hive.

The Warré hive uses top bars rather than full frames. When bees need more space, boxes are added below the existing ones. I do not have experience of this hive but it always strikes me that in a very good season this method of adding boxes would involve a great deal of effort.

Those wishing to pursue 'natural' or 'sustainable' beekeeping have a choice of hives, which include the Golden Hive, the Sun Hive, the Perone Hive and the Wild Hive. They are all workable designs but I would recommend that you don't try them until you have gained some experience of working with bees and have learned the first principles of how they operate. These hives require different management strategies and much of what you read in basic beekeeping books won't necessarily apply to them.

WOOD VERSUS PLASTIC HIVES

Most hives in the UK are made of wood, either western red cedar or pine. If you treat them properly with a preservative (making sure that it is one that is not fatal to insects!) or paint, they will last many years. If they do get damaged, they are easy to repair.

Hives made from high-density polystyrene are now available and increasing in popularity. They have the advantage of providing good insulation for the colony, especially in winter. The outer surfaces need to be painted with a water-based (masonry) paint to protect them from the effects of ultraviolet light. If they are damaged – and I have seen horrendous photos of one that attracted the attention of a woodpecker – then you will most likely have to buy a new one. However, hives can be protected from woodpeckers with small-mesh wire netting so, if you take care, you should not have this problem. These Styrofoam hives are easily damaged by hive tools and by rough handling.

Key idea

All the various hive designs available are suitable for a bee colony, so choose the one you wish to use and stick to it. Check locally to determine which hive is the most popular and therefore likely to be the most suitable for the local bees.

Remember this

Different hive designs are not interchangeable. Having different hives will severely limit your ability to solve problems between colonies.

Frame orientation: warm way and cold way

The way the frames are placed in the hive has nothing to do with the internal temperature but these names have just stuck. The bees control the temperature quite competently themselves. However, the orientation does influence the way the brood nest is arranged over the frames.

The brood nest is roughly spherical with an arc of pollen over it and then honey stores above that. It is found close behind the entrance to the hive. The frames in a hive cut through the sphere in slices. With frames running cold way, each will contain brood, pollen and honey, generally closer to the front of the hive. With frames oriented warm way, the first 'slice' behind the entrance will be through the pollen arc. Subsequent slices will show an oval patch of brood with the pollen arc above, right across the frame. Honey will be stored in the upper corners. The bees don't mind which way the combs run but it affects the way you approach the hive.

With frames the cold way, you need to stand at the side of the hive with the entrance on your left or right. Here you risk returning bees being upset by your presence and trying to sting you. Indeed, some bees are known as 'ankle biters' as they seem to prefer attacking at that level! With frames the warm way, you stand behind the hive, well away from the entrance with much less chance of upsetting the bees.

Being square, the National is the only hive that gives you the choice of the orientation of the frames without having to modify the standard design. Rectangular hives like the Langstroth can be worked warm way but the floor has to be modified to block the short side and provide an entrance on the long side.

Second-hand equipment

You are likely to find second-hand equipment advertised in the beekeeping magazines or available from members of the beekeeping association. This may look like a real bargain, but look carefully before you rush in to buy.

Firstly, make sure that the equipment on offer is compatible with your current hives. As we have seen above, parts of one design are generally not compatible with those of another. I strongly recommend that you use only one type of hive for your bees. If you have, say, a Langstroth and a National, you will be unable to swap frames and combs between them. If for some reason you lose the queen in one hive, you will be unable to insert a frame of eggs and young larvae from the other from which the original colony can raise a new queen. With all the same type of hive in your apiary, this operation will be easy.

Make sure that the equipment is in good order. Check that the boxes are square (and the right dimensions if they have been handmade) and do not have any gaps that stop the hive being bee-tight. The boxes need to be sound.

Key idea

Do not purchase second-hand comb unless it is in a hive containing bees. Comb carries disease. The spores of American foul brood can persist for over 80 years and still reinfect a colony. If you are offered frames with comb, cut the comb out and burn it. Then scrape the wax and propolis from the frames and boil them in a strong solution of washing soda (1 kg, or 2.2 lbs, of washing soda to 5 litres, or 1.3 US gallons, of warm water and a squirt of washing-up liquid to help clean off the propolis). You can then fit them with new foundation and reuse them.

CLEANING EQUIPMENT

Scrape any boxes clean and run a blowtorch over the inside surface of the wood to kill disease organisms. Catch the scrapings on newspaper and then burn them. Some hives are fitted with plastic runners rather than metal ones. Remove these before turning on your blowtorch! Heat the wood until it goes a uniform coffee-brown colour, but be careful not to set it on fire! Pay particular attention to the corners.

You obviously cannot use a blowtorch to sterilize a polystyrene hive but it can be scraped and then scrubbed with a washing soda solution as above.

Frames

Different frames fit the brood and super boxes of the various hive designs. You usually purchase them as separate parts and assemble them yourself. They come in packs of ten, which is rather inconvenient if you are using National hives, which take 11 frames. However, you will soon find a use for the spare ones! Your local association may well have a bulk order scheme for these and other equipment.

A frame consists of a top bar, two side bars and two bottom bars. The side bars fit at right angles to the top bar and the bottom bars slot into two grooves at the bottom of each side bar. A removable wedge in the top bar is used to secure the sheet of foundation in place. The groove on the inside face of the side bar also holds the foundation in place (Fig. 4.5).

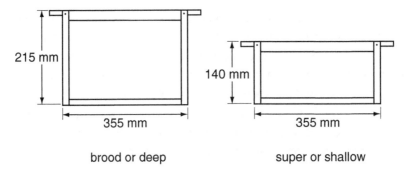

brood or deep   super or shallow

Figure 4.5 British Standard National frames

Side bars come in three main types:

1 **The DN1 (for brood frames) and the SN1 (for super frames)**
   (Fig. 4.6). This is a straight, narrow side bar. Spacers are
   fitted to the top bars to ensure that the frames sit at the
   correct distance apart. The commonest spacer is the 'plastic
   end', which slips over the lug. The end butts up to the spacer
   on the next frame. In the brood box, combs need to be two
   bee spaces wide, achieved with 'narrow' plastic ends. In the
   supers, 'wide' plastic ends can be used to increase the spacing
   so that the bees increase the depth (and volume) of the honey
   storage cells.

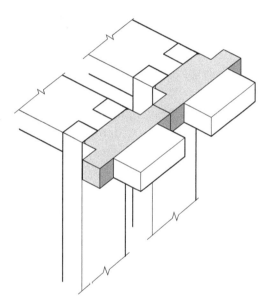

Figure 4.6 DN1 or SN1 frames with narrow spacers

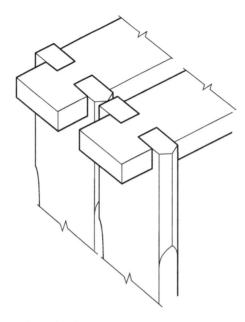

Figure 4.7 Hoffman self-spacing frames

2 **The Hoffman (Fig. 4.7).** This makes up a self-spacing frame because the top of the side bar widens out so that the combs are set the correct distance apart. The bottom of the side bar is the same width as the SN1 pattern. At the top, one edge of the side bar is flat and the other one is chamfered. When side bars are pushed together, there is only a narrow point of contact between adjacent frames, minimizing propolization. Put the grooves on the side bars facing inwards and the frames will match up properly. From experience, I would recommend using Hoffman frames in the brood box rather than plastic ends, but both systems will work.

3 **The Manley.** This side bar is a wide bar that is the same width throughout its length. Manley frames are generally used in supers and stop the frames swinging if the box is being moved, particularly on a vehicle.

The other way of spacing frames is to use castellations (Fig. 4.8). These are flat metal strips with square notches cut into them to accommodate the frame lugs. They are available with nine, ten or eleven slots depending on the hive type and whether you want narrow or wide frame spacing. They are nailed inside the box in the same way as the runners and adjusted so that the frames are correctly positioned either with a top or bottom bee space, as appropriate. If they are not in the right place, the bees will fill the gaps with either propolis or brace comb. Using castellations in the brood box makes manipulations more awkward and I would recommend using them only in the supers.

Figure 4.8 A ten-slot castellation

The type of super frame you use will affect the type of extractor you can use. SN1 frames will go into both a tangential and a radial extractor but, if you put Hoffman or Manley frames in a tangential extractor, there is a greater chance of the comb breaking because it is not held up against the extractor cage (more of this later).

Just to complicate things, top and bottom bars come in both narrow (standard) and wide versions. Although the initial cost will be slightly higher, I would recommend using the wide version of each. This cuts down the amount of brace comb between the combs and makes inspections and manipulations easier.

Remember this

Acquire enough equipment sufficient for all the colonies you have or anticipate having during the season. This should include four supers per colony. Acquire sufficient equipment to undertake swarm control manipulations on all of your colonies at the same time.

MAKING UP FRAMES

It is very important to make up your frames accurately. If they are not square and flat, the bee space will be compromised, making life more difficult than necessary. You can make a jig that ensures all the parts go together at right angles or, alternatively, you can check each one with a set square. It is also essential that the frame is not twisted in the vertical plane, so check it by eye and adjust if necessary.

Before you start nailing up frames, assemble the necessary tools: a hammer, a craft knife, a jig or set square and a pair of pliers (for removing bent nails). You will also need frame nails, available from beekeeping suppliers.

I would first push a frame together so that you can see how everything fits. Make sure that the grooves on the side bars face inwards to support the foundation. With Hoffman frames, assembling them correctly ensures that the vertical edges butt up correctly when they go in the hive.

1 First, take the top bar and remove the wedge that is attached with a thin sliver of wood. Shave off any excess sliver with a craft knife. Put the wedge somewhere safe, as you will need it later when you insert the foundation.

2 Then push the two side bars on to the top bar. Check that they are at right angles with the set square and secure them in place, driving a nail through into the top bar from each side.

3 Push one of the bottom bars into the grooves in the side bars on the opposite side from the missing wedge. Adjust it so the ends are flush with the outsides of the side bars. The bottom bars make the frame rigid and hold the foundation in place vertically. They do not carry any weight.

4 Insert a nail at each end of the bottom bar from underneath so that it points up the length of the side bar rather than across it. This makes it easier to remove the bottom bars when you need to replace the foundation.

Frames can be made up at any time and stored in a brood or super box, but make sure that you don't mislay the wedges. It is better not to fit foundation until shortly before you want to put the frames into your hive. Wax is malleable and sheets of foundation can warp easily. They cannot then be flattened out again and, if you give them to your bees, they will draw wavy comb. This isn't a problem for them because they will ensure that the faces of adjacent combs are the correct distance apart, but it will make your manipulations more difficult, especially if you want to move a frame from one position in the box to another or even into a different box because the profile will not fit with the other combs.

Foundation

Bees are quite happy building comb in an empty space, as they have done in cavities for millennia. If you put empty frames in a hive, they will do just that, but not necessarily within the frames. Building it across the frames may be fine for the bees, but not for the beekeeper. The idea is to persuade the bees to build comb where you want it, neatly within the frames, which you can then remove when you wish. The solution is to put a sheet of foundation into the frame.

Foundation is generally made of beeswax, although plastic versions are available and widely used in the USA. The hexagonal cell pattern is embossed on the surface on each side with cell bases offset, just as in the natural situation. When you give foundation to bees, they use it as the basis for building the cells of their comb within the frame, just as you wanted.

WIRED FOUNDATION

As we have seen, beeswax can be fragile. In the UK, foundation is generally sold as 'wired' to make it stronger when used in the brood box and less likely to collapse in the super frames when they are put into the extractor. The commonest pattern is a zigzag of wire, embedded into the wax, running up and down the sheet with larger loops left outside the sheet at the top edge and smaller ones at the bottom (Fig. 4.9). Wired foundation also comes with six or seven crimped wires embedded vertically into the sheet, with the ends sticking out slightly at each side.

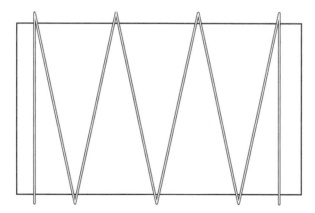

Figure 4.9 Wired foundation

UNWIRED FOUNDATION

If you want to produce 'cut comb' honey, where sections of the entire comb are cut out and eaten, you must use unwired foundation. For cut comb production, 'thin' foundation is available, reducing the amount of wax in the midrib and making the product more palatable. Sheets of standard unwired foundation can be used to make rolled beeswax candles.

You can make your own foundation but it is much easier to buy it direct from the manufacturers. The larger equipment suppliers offer a part-exchange scheme where you give them your cleaned beeswax and they give you the equivalent weight in sheets of foundation. You won't be able to do this until you have collected some wax from cappings or from combs that you have melted down for some reason.

STORING FOUNDATION

If you are not going to use it immediately, store your foundation flat. Place it on a flat board slightly bigger than its dimensions, putting a similar board on top. A small weight on top will help, as will keeping it in a relatively warm (but not too warm) place. Too cold and the wax will go brittle. Too hot and it will melt. The warm, constant temperature in an airing cupboard is excellent – but make sure that you get permission first!

Key idea

Keep foundation flat until it is required.

RENEWING FOUNDATION

It is good practice to renew foundation every two to three years to reduce the risk of harbouring disease. Plastic foundation is usually supplied already fitted in a one-piece plastic frame. The embossed plastic sheet has a thin coating to make it more attractive to the bees. Such frames can be sterilized and recoated if necessary. You can also replace comb in wooden frames.

To renew foundation, first remove the old comb. Wrap it up to stop the bees finding it and then dispose of it. It can be burnt and any residue put on the compost heap. Scrape the top bar clean, then remove the wedge and clean out any wax fragments. Scrape the wax off the side bars and clean out the grooves (there is a handy tool for this available from suppliers). Remove one of the bottom bars and scrape both of them clean.

Frames should be sterilized before reuse to get rid of any disease organisms lurking in the crevices. Use a blowtorch to lightly scorch the surfaces or scrub them clean with a strong solution of washing soda (1 kg, or 2.2 lbs, of washing soda to 5 litres, or 1.3 US gallons, of water). Sets of frames can be boiled in a washing soda solution in a 'Baby Burco'-type boiler to clean and sterilize them. After cleaning, you can insert clean sheets of foundation into the frames and give them back to your bees.

Remember, be very careful if you purchase drawn comb. Replace this with new foundation after you have sterilized the frame parts.

Remember this

Drawn comb is a very valuable commodity in a number of beekeeping operations. Take care of it.

FITTING FOUNDATION INTO THE FRAME

Although purchased foundation is cut to a standard size, you may have to shave a little off to make the sheet slide into the frame.

This is how to fit wired foundation:

1 Slide the sheet of foundation up the grooves in the side bars, leading with the larger loops/wires.

2 Bend the loops/wires over at right angles so that they fit snugly into the rebate in the top bar from where you removed the wedge.

3 Replace the wedge so that it covers the wire loops and the top part of the foundation.

4 Use three frame nails to secure the wedge in place. You can drive them through the wire loops , though this isn't essential.

5 Finally, put the second bottom bar in place, with the foundation in the gap, and nail it in place as before.

Making up a hive

Hives can be purchased assembled and ready to go. This is certainly the quickest way to get started but probably one of the most expensive. Most beekeepers buy their hives 'in the flat' and put them together themselves. The pack contains all the necessary pieces, sawn to size and shape for your chosen design. Assembly is generally straightforward, although the Modified National needs a little more thought because of the fillets on two of the sides. You can make your own hives but you will need to be fairly skilled at woodwork to do this because the bee space must be correct everywhere. Plans can be downloaded from the Beesource beekeeping website at www.beesource.com/build-it-yourself/

However you make your hive, the essentials are that the internal dimensions are correct and that the boxes are square. All the boxes must fit together in any combination without any gaps, particularly with single-walled hives. The sidewalls must be flush at the corners. You can use a jig or long sash cramps to hold the box square before you nail it together. You will minimize the risk of the wood splitting if you pre-drill holes for the nails at the corners. It is worth strengthening the joints by applying waterproof glue before you nail them, because your hive will take a lot of knocks during its long life.

It is important to position the runners or castellations correctly, to maintain the bee space. In the National hive, they go on the sides with the fillets. In other designs they fit into the rebates on two opposite walls. Use a frame to help get them in the right position.

To assemble a National hive, take these steps:

1   Construct the two walls carrying the fillets first and then join them to the other two sides. Make sure that the bottom fillet is correctly positioned so that there is a bee space under the bottom side of the wall.

2   Set the runner in place but do not fix it.

3 Put a frame at one end of the box and adjust the runner or castellation for top or bottom bee space, as required, before nailing it to the sidewall. Don't knock the nail all the way in at this stage as you may have to adjust it when you move the frame to the other end of the box and repeat the process.

4 When the runner/castellation is in the right place, nail it firmly to the wall. Repeat this for the opposite wall.

Your floor, inner cover and roof must also be square. Open-mesh floors prevent varroa mites from re-entering the hive but they also stop house bees clearing out debris that falls through the mesh. This can be a breeding ground for wax moths, so you need to clean the trays regularly, even if you are not monitoring mite levels. Cover the ventilation holes in the roof with mesh to prevent unwanted visitors. To protect the roof, cover it with a sheet of metal or waterproof material, bending the edges over the sides and nailing them in place.

Finally, protect the wood with a coat of preservative, making sure it is one that is not harmful to insects. An alternative is to use hive paint available from equipment suppliers. Paint polystyrene hives with a water-based masonry paint to protect the material from deterioration by ultraviolet light.

 Try it now

* Learn about the different parts of the hive and how the hive goes together.
* Appreciate the importance of the bee space within the hive.
* Choose one hive type and stick to it so that frames and boxes within the apiary are compatible.
* Be careful when acquiring second-hand equipment because of the possibility of spreading disease.
* Make up frames and hive boxes before they are needed but don't fit foundation until just before frames are given to the bees.

Focus points

* Decide on the hive design you are going to use before you acquire your bees.
* Look at a range of hive designs and talk to other beekeepers if possible before making your choice.
* Make up frames and hive boxes very carefully so as to maintain the bee space and make your beekeeping life much simpler.
* Do not buy second-hand comb if it is not being used by bees. If you do acquire some, cut out the comb and burn it before cleaning and sterilizing the frames and fitting new foundation when needed.

Next step

The next chapter will look at the essential tools you will need to inspect a beehive. The minimum is a hive tool and a smoker. Suitable fuels are described together with the process of lighting the smoker. The importance of keeping records of your apiary visits and colony inspections is stressed.

5

Essential equipment and records

To inspect your bees easily and successfully, you will need certain tools. There are many gadgets on the market, but the two absolute essentials are the hive tool and the smoker. Probably the next most important is a set of record cards, one for each colony, to enable you to keep track of your apiary visits and inspections.

The hive tool

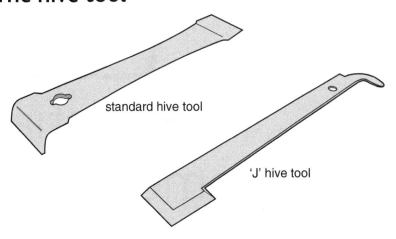
standard hive tool

'J' hive tool

Figure 5.1 Standard and 'J'-type hive tools

The hive tool (Fig. 5.1) is used for separating boxes and for freeing the frames so that you can lift them from the hive. It is also used to scrape beeswax and propolis from top bars, boxes and queen excluders.

There are two basic designs – the standard hive tool and the 'J' pattern. Try them both and see which you prefer. I started with a standard one but now prefer the 'J' pattern.

▶ **The standard hive tool** is a flat piece of metal, usually slimmer in the middle to make it more comfortable to hold, with a flat edge at one end and a curved edge at the other. Both ends are chamfered so that they can be used as scrapers.

▶ **The 'J' design** also has a flat, chamfered end but the other one is curved over in the same plane – hence the name.

The flat end of each design is used to prise hive boxes apart. The curved end of the 'J' tool is useful for lifting the frame lugs so that you can grasp them to remove the frame from the box. This is particularly true for the first frame out of the hive. With the standard hive tool, you will probably find this manoeuvre easiest if you use the curved end, but try it and see what suits you best.

Whichever hive tool you choose, you need to learn to keep it in your hand all the time you are inspecting your bees. You will use it frequently and you don't want to waste time picking it up and putting it down – or looking for it. I know a beekeeper who attaches his hive tool to a bungee cord attached to his belt. You might find this useful because you can retrieve it easily if you let it go.

Keep your hive tool clean. You will be using it on different hives and it is therefore a potential carrier of disease between them and particularly between apiaries if you end up keeping your bees at more than one location. Before going to the apiary, scrape off the wax and propolis and clean the hive tool with a proprietary kitchen surface cleaner. You can also scrub it with a strong solution of washing soda, such as that used for disinfecting frames. Take a bucket of this solution to the apiary with you and use it to disinfect your hive tool between colonies.

Remember this

Learn to keep your hive tool in your hand throughout manipulations.

The smoker

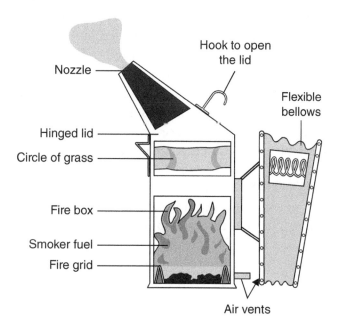

Figure 5.2 The smoker's construction

The second essential piece of equipment is the smoker (Fig. 5.2). You will use this to move bees away from areas you wish to inspect and to calm them down if they start getting too interested in what you are doing.

Smokers come in different sizes and my recommendation is to buy the largest one you can use comfortably. There is nothing more annoying than your smoker going out just when you need it most. The fuel can be topped up but the larger the barrel, the less often this is required. A larger smoker is also easier to keep alight.

The main compartment of the smoker is the firebox or barrel made from galvanized steel, copper or stainless steel. Some designs incorporate a heat guard. There is a grate at the bottom of the barrel, which supports the fuel and enables air to be blown up through it to keep it alight. Spring-loaded bellows are attached to direct the air into the barrel through a hole at the base, opposite a small tube welded around a hole in the base of the barrel. The barrel is closed at the top with a nozzle that emits the smoke in the direction required.

There are various potential fuels but you need one which produces a cool smoke so as not to roast your bees. One of the best fuels is dry, rotten wood or well-rotted jute sacking. These may not be readily available and you need to be sure that the sacking has not been treated with a chemical harmful to bees. Other fuels that work well include dry leylandii leaves, dry pine needles and machine planer shavings, such as those sold for animal bedding. You can also use smoker pellets or raw cotton waste, available from equipment suppliers.

LIGHTING A SMOKER
Lighting a smoker and keeping it alight is not always as easy as it sounds. The real test is to put a lighted smoker to one side and return some 30 minutes later to find our whether it is still alight and ready to use. In America, beekeepers hold competitions to see whose smoker will stay alight the longest.

Here's how to light a smoker using planer shavings:

1 Take some newspaper and tear off about a quarter of a broadsheet piece or half of a tabloid. Crumple it loosely, hold it over the open smoker barrel and light the bottom. Puff the bellows gently as you lower the paper inside.

2 Continue to puff gently and sprinkle in some shavings. At this stage you need to see the flames, so don't put in too many shavings at once. Continue puffing and adding fuel so that you build up a bed of red-hot embers.

3 When the shavings are well alight, add more to fill the barrel but don't press them down as this may extinguish them or prevent the air from passing through. Your smoker will be producing a lot of smoke.

4 If you leave it for ten minutes and find it springs back into life when you return, you know you have cracked it!

If you are using a loose fuel like shavings or leylandii leaves, take a handful of long grass, twist it into a circle and place it on top of the fuel. This will stop bits of smouldering fuel from being blown out of the nozzle.

If you have access to rotten wood, light a small piece and put this into the barrel in place of the newspaper. It will burn for longer and help to ignite the other fuel.

Remember this

Practise lighting your smoker and keeping it alight. Try out different smoker fuels and collect together the kind that suits you best.

EXTINGUISHING THE SMOKER

Almost as important as lighting the smoker is extinguishing it safely. Do this by depriving the fuel of oxygen. Block the nozzle, either with a solid bung (a flanged cork works well) or a twist of grass. Covering the hole at the base of the barrel, say by stuffing grass or a paper handkerchief in the gap, speeds up the process. Laying the smoker on its side will also put it out more quickly.

Although the smoker may have been extinguished, the barrel will remain hot for a while. I know a beekeeper who thought his smoker had gone out. He put it into the back of his vehicle. The rubber mat still bears the mark. Fortunately, he realized what was happening before more damage was done. Some beekeepers carry a tin box in their vehicles to prevent such accidents.

Record keeping

I urge you to get into the habit of keeping records of your beekeeping activities right from the start. This needs to become second nature. When you start beekeeping and have only a couple of colonies, you will probably remember what you did and when. However, as time goes on, and especially if you increase the size of your apiary, remembering gets more difficult.

Records don't just remind you of what you did on the last inspection or what you need to do on the next one; they are essential when you want to look back over the year. They will show which colony collected the most honey, which tried to swarm and which was the best, or worst, tempered.

You can record anything you think will be useful. There is no set format for a record card, although several ready-made versions are available. These include one produced by Bee Craft Limited (www.bee-craft.com/shop), which has the advantage that it is part of a series of laminated apiary guides that all fit into a custom-sized folder. The apiary guides include sets on bee diseases, the varroa mite, integrated pest management, swarming and swarm control. They provide aide-memoires for use in the apiary and are much more portable than a large beekeeping book.

Some beekeepers number each set of hive boxes. It took me a while, but eventually I came to appreciate that it is much better to associate the number with the queen. If she swarms, for instance, and ends up in a different box, the number simply moves with her. After all, she is the mother of all the bees in the colony and responsible for the colony's characteristics whichever hive she is in, and this is what you need to keep track of. I pin a plastic label with the number on the side of the hive. If the queen moves to a different box, I move the label as well. You do not have to physically number the queen, although you can get small numbered discs that can be attached to her thorax with a dab of paint or glue if you wish.

In due course, you will want to look back and assess the queen's and her colony's characteristics to plan your future colony management. If your bees are particularly defensive (or aggressive from our point of view), or if they follow you back up the garden after you have finished your inspection, you will probably plan to replace the queen with one that produces more docile workers in order to improve the colony's characteristics as offspring from the new queen become the majority.

Remember this

* Keep your hive records together so that you can refer back to them at the end of the season.
* Number the queen rather than the box she occupies. The records refer to the activities of her colony and they therefore need to go with her if she is transferred to another box.

Whatever system you decide to use, remember to keep your records up to date. You may simply start by keeping a diary but in the end you will probably find it easier and quicker to maintain separate records for each queen.

Computerized record systems are available but you will probably want to consider these only if you have a larger number of hives. Some beekeepers use a small recorder for making their comments while in the apiary and then transpose these on to a paper record when they get back home.

The method isn't important. What is important is that you record the condition of your colony, your actions and what you need to do next time you visit the apiary.

Key idea

You may find a bee box useful – for your bits and pieces rather than for your bees. This keep things together and reduces the chances of arriving at the apiary to find that you have left something vital, such as the hive tool or the matches, behind. This isn't so bad if your hives are in your garden but it is a bit more annoying if you have just climbed up to the roof or driven to the allotments.

The bee box needs to be sturdy and, if you going to keep your smoker in it, made from a fire-resistant material. It needs a suitable handle, making it easy to carry. Divisions inside will help keep your things in order.

Try it now

* Choose the hive tool you find most comfortable to use.
* Choose your smoker and acquire the largest one that you can operate easily.
* Decide how you are going to keep your records before you start your inspections.
* Fit out a bee box to carry your kit, especially if your apiary is not close to home.

Focus points

* Your two primary essential tools are your hive tool and your smoker.
* Choose the design of each that suits you. Try them out first if possible.
* Get into the habit of keeping records right from the start.

Next step

The next chapter looks at ways to minimize your exposure to bee stings and the clothing you need to protect you from potential stings when you open a colony. Primarily, you need a veil. Suitable footwear is also important to avoid slipping in the apiary. As a beginner, you will probably also wish to wear gloves.

6

Protective clothing

Bees defend their nests by flying round their attacker, emitting a high-pitched buzz. This is the warning, but if the potential predator takes no notice, the bees bring out their ultimate weapon – the sting.

At the very least, being stung is painful. It leads to swelling and irritation until (and if) you become immune to the effects of the venom. In severe cases, someone very allergic may suffer anaphylactic shock and even die. I am not trying to be melodramatic and paint the worst-case horror scenario, but this is something that you need to be fully aware of before you start beekeeping.

Fortunately, there are several ways you can minimize your exposure to bee stings.

Minimizing stings

The vast majority of beekeepers will eventually become immune to bee venom, and future stings will just be painful and inconvenient. If you find your reaction gets worse each time you are stung, you should consult your doctor. Courses of treatment are available that build up your immunity safely, although this may not happen quickly or even at all. It is better to undertake treatment in a safe, controlled environment than suffer a sudden reaction that requires urgent emergency treatment.

Courses are offered which give training in the steps required if someone at the apiary suffers an extreme reaction, and your local beekeeping association should be able to arrange to hold one of these. I recommend that you attend.

Prevention is better than cure so let's look at the steps you can take to minimize the chances of being stung.

GOOD-TEMPERED BEES

One of the first and best ways of minimizing the chance of being stung is to keep only good-tempered bees. This is important, not only for you, but also for others in an urban environment. Some bees will defend only a small area around their nest entrance. Others will patrol a much wider area and these are the bees that will cause problems with your neighbours or passers-by.

I once went to an apiary meeting where we were met at the entrance by guard bees from the hives. They buzzed around all the while the hives were being inspected. They then followed us for quite a distance as we left to go for tea in the village hall down the road. Some 45 minutes later, as we emerged to go home, some were even waiting for us in the street!

Admittedly, this is an extreme case, but there is no need to keep bad-tempered bees and certainly not when you live in close proximity to non-beekeepers. Surprisingly, they may very well not share your love of these insects and many people are frightened when a bee comes anywhere near them. It is your responsibility not to put people in this position.

Protective clothing

Figure 6.1 Be keepers in protective clothing collecting a swarm

Even good-tempered bees will sting on occasion, so you need to wear protective clothing (Fig. 6.1). At the very least, this is a veil to protect your face. Bees are attracted to contrast and movement – both of which occur around the eye area. Being stung on the eye is dangerous. At the best, the area will swell, sometimes spectacularly. At worst, you may damage the eye itself.

THE VEIL

Veils come in a wide range of shapes and designs. I started with the very simplest 'hat and veil', which consisted of a cylinder of black cotton net supported by the hat's wide brim. Black net allows you to see out more easily than white net. The other end of the net was wide enough to sit over my shoulders and be held down by tapes under my armpits. The Rolls-Royce version has a ring of wire sewn into the net around chin level to keep the net away from your face. If you use one of these veils, wear it with the bottom edge of the net on top of your shoulders rather than over the top of the arms, which leaves a gap through which a bee can get inside.

This type of veil is usually worn with a cotton boiler suit, which protects the rest of the body from stings and helps keep your clothes clean. Propolis is a very sticky substance that is difficult to remove from cloth. If you do get some on your clothes, try a proprietary 'tar and oil' stain remover, but test it first to make sure that it won't leave a mark.

Some 40 years ago, Brian Sherriff designed and manufactured a new type of veil and now all veils of this type are known as 'Sherriff' veils. The new concept was of an all-in-one veil where the hood was attached to, initially, a sleeved top and then to a full boiler suit. A zip round the base of the hood enables it to be thrown back when not in use and removed completely when the suit is washed. The back of the hood is made from cloth. The front consists of black mesh supported by two semicircular arches to keep it away from the face.

If you can attend one of the major beekeeping events that include a trade show, you can look and try the different veils on offer and choose the one that suits you best. If you can't do this, talk to members of your local association and have a look at the veils they use.

In some ways, the Sherriff-type veil is too good! If offers you excellent protection from the bees' attentions. However, what it can't do is protect your neighbours. You may be blissfully unaware of the attention they are getting because you are safe and secure and absorbed in the task in hand. Try to be aware of what is going on more generally when you are inspecting your colonies.

Remember this

Your veil must be bee-proof for you to be safe in the apiary. Wear your veil whenever you open a hive so that your face and particularly your eyes are protected from stings.

GLOVES

Certainly as a beginner, you will want to wear gloves to protect your hands. This is the part of the body that gets closest to the bees and is therefore particularly vulnerable. Again, you have a choice.

My personal preference is for light leather gloves with cloth gauntlets that cover the forearms. However, you can wear something as simple as a pair of rubber washing-up gloves. Some beekeepers wear thin latex gloves over leather ones. Rubber gloves are cheap and can be discarded when they get dirty. With disposable gloves, you can change them between apiaries or even between hives to reduce the chance of spreading disease.

Remember this

Thick gloves can make manipulations clumsy and this can upset the bees. Make sure that you can handle frames smoothly when wearing your chosen gloves.

FOOTWEAR

Some bees are 'ankle biters'. Wearing rubber boots not only protects you from such bees but also gives you a better grip so that you don't slip. You don't want to fall over when you are carrying a heavy box of frames!

Key ideas

* Bees will sometimes sting for no apparent reason, so you must make sure that you are protected when you open a colony.
* Veils come in different shapes and sizes – and colours. Choose the one that is most comfortable.
* Rubber boots provide protection from stings to the ankle area and also a good grip on the ground in the apiary.

Try it now

* Obtain your personal bee-proof protective equipment.
* Try on as many different veil designs as possible to determine the one in which you feel confident and comfortable.
* Check the bees you wish to acquire and only keep colonies that are good tempered.
* Try different types of gloves to determine which type you wish to wear.

Focus points

* Modern veils are very effective at protecting you from your bees, but they will not protect your neighbours.
* Gloves can be a means of spreading diseases between colonies. Wearing disposable gloves over another pair means that you can minimize the risk.
* You need to make sure that your footing is stable and safe in the apiary, so wear suitable footwear.

Next step

The next chapter describes the process of acquiring your bees. We discuss the best time of year to buy them, what sort of bees to choose, and what to look for in a colony and how to make sure that the bees are healthy. We also give advice on how to transport them and transfer them to your hive.

7

Acquiring your bees

You have your apiary site, you have chosen your hive type and your personal equipment, so all that is left now is to choose your bees. After all, that is surely what beekeeping is all about? Even though keeping bees is not the same as keeping chickens or goats that are generally contained in a field or a pen, choosing bees is a little more complicated than it sounds.

Once you become a beekeeper, you are committed to do all you can to help your bees survive and thrive. You are taking on responsibility for living creatures. This is why, just as when keeping any sort of livestock or pet, it is important to learn about them and prepare properly so that you can look after them as best you can.

When to buy bees

Most beginner beekeepers are desperate to acquire their bees and this is totally understandable. I bought my first colony in the autumn from the tutor of the beginners' course I was attending. Looking back, this probably wasn't the best move as I had only been to half of the theory classes and had no experience of actually handling bees. The two practical classes included in the course were scheduled for the beginning of April. That winter was particularly bad and halfway through I had to ask a beekeeper friend to come and check that the bees were OK. As it happens they were, but it would have been better all round if I had waited until late spring to acquire my first colony.

We tend to think of winter as being the time when colonies are most at risk, but it is actually spring when they are at their most vulnerable. The queen is laying more eggs and there are many more mouths to feed. If the weather prevents foragers collecting fresh pollen and nectar, the colony is totally dependent on its remaining stores and access to them. Colonies are more likely to die of starvation in spring than through the winter. I would therefore recommend that you resist the urge and do not get your bees until you know that the colony has come through winter successfully.

What sort of bees?

What sort of bees are you looking for? 'Surely,' you say, 'a bee is a bee is a bee.' In one sense that is true because individual bees and colonies follow the same life cycle and development patterns, as we have already considered. However, different colonies can exhibit very different characteristics, primarily because of their genetic make-up.

In an urban situation, the type of bee you keep is possibly one of the most important decisions you will make. The queen largely determines the colony's character, but so do the drones with which she mates. Some colonies are very calm and defend their nests only under extreme provocation. Others are quick off the mark and, as we have seen, will even come to the apiary entrance to warn you not to mess with them. Which would you rather have? And which would your neighbours want you to keep? That's not a difficult question to answer!

Some beekeepers will try to tell you that bad-tempered bees collect more honey. Don't believe them. I have known good-tempered bees that produced a great honey crop and bad-tempered bees that harvested nothing. I have also known the reverse.

Key idea

The bees you keep should be good tempered and make it a pleasure to look after them.

As well as temper, the other important characteristic is the colony's propensity to swarm. There are ways of preventing and controlling swarming but these involve you recognizing the signs of swarming preparations and then taking the necessary action at the right time. These have to be carried out regardless of the weather or other things in your life because they are linked to the development of the new queens. Since bees are living creatures, there will be times when even the most careful of swarm control procedures will fail and a swarm will leave the hive.

Swarming, its prevention and control are covered in Chapters 9 and 10 and you are encouraged to study and understand what is going on. At the mention of swarm control, many beekeepers' eyes glaze over and their brains go numb. Swarming and swarm control are not difficult to understand. You just need to grasp the principles behind them and realize how they are linked to the life cycles of the queen and worker and the natural process within the colony itself.

LOCAL OR IMPORTED?

I would strongly recommend you look for local bees that show the characteristics you desire. Such bees are acclimatized to the local conditions such as the weather patterns and available forage.

Over the British Isles, for instance, the seasons differ in different areas. Spring comes early in Cornwall but later as you move north. Colonies in the north of England will start expanding their brood nests later than in the south and will also start preparing for winter sooner.

The climate in urban areas also differs from that in the countryside. Towns are a few degrees warmer than rural areas, which means that flowers will begin to bloom earlier. Once the oilseed rape has finished flowering, there is often very little on which rural bees can forage, while in towns a continuous succession of suitable flowers is much more likely.

What you need are bees that are in tune with their local environment. If they are of the type that uses all incoming nectar to rear brood and there is a reduction in the available forage, you will have to provide supplementary food. Locally adapted bees will regulate their brood rearing to local conditions.

We all know that the 'grass is greener on the other side of the fence'. However, this is not necessarily so in practice. Look for bees that are used to how things happen where you are.

You may be tempted to buy a queen or bees of a different race or ones imported from a different country. Before you do, consider two things. If these bees are used to living somewhere where food is available through most of the year, they will not do well if you introduce them to a region where the seasons are distinct and short.

Bee genetics also comes into this. Virgin queens mate on the wing with free-flying drones from any colony in the area. Unless you use instrumental insemination, where you can choose the drones that donate sperm to the queen, you have no control over the matings. Research has shown that, when you introduce an 'exotic' bee into your area, virgin queens produced as part of the swarming impulse mating with the local drones leads to an increase in bad temper. This may not be in the first generation but it will certainly arise with further out-crosses.

The genetics of local bees will be much more similar to those of the local drones and you reduce the chances of bad temper occurring. It doesn't mean that it won't but the risk is diminished.

I do not apologize for stressing once again the importance of considering your neighbours. This really is vital in an urban or semi-urban situation. Actually, it is important in rural situations, too, but it's just that there the neighbours are generally not so close. In any case, keeping bad-tempered bees is simply unpleasant and becomes a chore.

Remember this

The temper of a colony is no guarantee of a good honey crop, but beekeeping is a much more pleasant occupation if you have good-tempered bees. Introducing bees of a different race or from a different country will lead to a deterioration in colony temper in due course.

What to look for when buying bees

There are several ways of acquiring your bees. The best is probably by purchasing a nucleus – a small colony where bees occupy three or four frames. This is easier to inspect and look after and, as your nucleus expands, your experience will grow with it.

The British Beekeepers' Association (BBKA) has a code of practice detailing what you should look for. These are the main criteria:

▶ The nucleus should contain bees, brood, food and a queen of the current or previous season, reared in the UK.

▶ The queen should have produced all the brood in the nucleus.

▶ She should be marked with the universal colour for the year in which she was raised.

▶ The seller should clip one of the queen's wings at your request if this has not been done already. This unbalances the queen so that she cannot fly properly and will not be able to leave with a swarm. This delays the swarming process but it cannot be used as a complete method of swarm control.

▶ The nucleus should contain the number of frames the seller states.

- At least half the total comb area should contain brood in all stages (eggs, larvae and sealed brood) with a minimum of 30 per cent being sealed and 15 per cent being drone brood (depending on the time of year).

- There should be no active queen cells at any stage of development.

- The frames should be well covered with bees, with a good balance between young house bees and older flying/foraging bees.

- The bees should be good tempered when handled by a competent beekeeper in suitable conditions. Clumsy handling through inexperience can cause the bees to react adversely.

- The brood and adult bees should show no signs of disease in any stage.

- The seller should provide information on any disease treatments undertaken, including that for varroa, and when they were administered. All treatments given must be legal as, once you have purchased the bees, you become responsible for any illegal substance that may be found in the colony.

Key idea

A nucleus you wish to acquire should be disease free and conform, in the UK, to the British Beekeepers' Association code of practice.

If there is any doubt on any of these points, I advise you to look elsewhere. If you are not confident to inspect the nucleus on your own, ask an experienced beekeeper to come with you. The seller should have no objections.

When your nucleus arrives, it should be ready to expand. This means there are sufficient bees to look after a good amount of sealed brood which is just about to hatch. The resulting new bees will augment the adult bee population as nurse and house bees, freeing up older bees to go out foraging. The age of the queen will also affect the speed of expansion as older queens have lower egg-laying rates and hence the nucleus will grow more slowly.

CHECKING FOR DISEASE

One of the most important checks is to make sure that your bees are free from the major bee diseases. Virtually all colonies in the British Isles are now infested with the varroa mite but treatment should have been given to reduce mite levels as low as possible to limit the mites' effects. Diseases are covered in more detail later, but the four notifiable diseases are American foul brood, European foul brood, small hive beetle and the *Tropilaelaps* mite. The last two have not yet been identified in the UK and with any luck that situation will continue. However, small hive beetle was positively identified in colonies in southern Italy in 2014 and so is moving closer to the UK.

Notifiable diseases are those that must, *by law*, be reported to the authorities, even if you only suspect their presence. The relevant authority depends on where you live. In England and Wales it is the National Bee Unit in York; in Scotland it is the local area office of the Scottish Government Rural Payments Inspections Directorate (SGRPID). In Northern Ireland get in touch with the Department of Agricultural and Rural Development and with Teagasc in the Republic of Ireland. See Chapter 23 for contact details.

It is very important that your bees are healthy. As a friend of mine says, 'Dead bees collect no honey!' Bees suffering from disease don't do well either. If you have any doubts at all about the health of a nucleus you plan to buy, in England and Wales you can ask the Regional Bee Inspector (RBI) to advise you. The National Bee Unit (NBU) employs RBIs and their teams of seasonal bee inspectors (SBIs) to inspect colonies for statutory notifiable diseases. All the bee inspectors are experienced beekeepers who are very willing to offer help and advice on all aspects of beekeeping, not just diseases. This is a free service in England and Wales and contact details can be found on the NBU website (www.nationalbeeunit.com). If you live in Scotland or Ireland, contact the relevant authority or ask an experienced local beekeeper to help you. In the United States, contact the US Department of Agriculture.

Transporting your bees

You will almost certainly have to collect your nucleus from the supplier, so buying locally helps reduce the stress the bees are subjected to during transport. They may be in a standard nucleus hive, a travelling box or a temporary container. Make sure that you know what, if any, of the equipment needs to be returned to the supplier. Some may ask for the equivalent in new items but, more usually, you will buy the box and frames along with the bees.

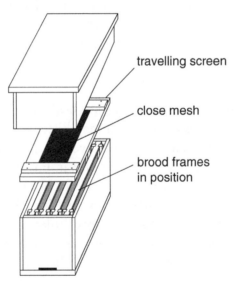

travelling screen

close mesh

brood frames
in position

Figure 7.1 A nucleus box with a travelling screen

When bees are confined to the hive with the entrance blocked, they have no means of cooling the cavity. You therefore need to replace the inner cover with a bee-proof mesh travelling screen with the same dimensions as the box and remove the roof in order to supply ventilation (Fig. 7.1). You don't want the bees to escape during the journey, so strap the hive securely and close the entrance. Hive straps are available from equipment suppliers. They can be tensioned so that hive parts do not come apart even in the event of an accident. The easiest method I know for closing the entrance is to stuff the gap with foam rubber. (I get ours from discarded cushions.) The foam can be cut into suitable lengths and pushed into the entrance with the hive tool. Timing the move for early morning or later in the day also means the journey will be cooler for the bees.

You don't want the box moving around in your vehicle, so wedge it securely in place. Don't leave your bees in a hot car during the journey. If a colony gets overheated, the comb can collapse and most of the bees will be drowned in the honey or squashed. You will have not only lost your precious colony but clearing up will be a very messy business. Keep your bees cool by spraying or dribbling a little (not too much) water over the screen.

Having got your bees home, place the nucleus on the prepared stand with the entrance facing the way you wish. Remove the entrance closure and allow the bees to fly. Sometimes they pour out of the entrance and other times nothing happens immediately and you fear that the colony has died. Wait for a bit and they will soon begin investigating their new home area. They will fly out in ever-increasing circles as they learn the landmarks around their hive.

Put the roof over the travelling screen to protect against the weather.

Even if your nucleus contains a lot of bees, they are likely to be OK in the nucleus box for a couple of days. It is best to let them get used to their new location, but in due course you need to transfer them to a full-sized hive.

Key idea

Remember that bees can overheat when confined to the hive for transportation, so provide bee-proof ventilation.

Transferring bees to the hive

First, make sure that you have all the hive parts to hand – floor, brood box, inner cover and roof at this stage. You will also need sufficient frames to make up the full complement of your chosen hive design, bearing in mind that you will be transferring over the frames from your nucleus. These frames are most likely to be fitted with foundation, particularly if you are beginner.

Try to choose a day when your activities are least likely to affect your neighbours. You will have a lot of bees flying around, which can be alarming to a non-beekeeper. Working carefully will minimize any problems:

1 Light your smoker and put on your veil. You will need your hive tool and will probably want to put on your gloves, particularly if this is the first time you have actually handled bees. Gloves will give you more confidence because you won't be distracted wondering whether the bees landing on your hands are going to sting you (they most likely will not do so).

2 Gently puff some smoke into the nucleus entrance to calm the bees. With a small colony and docile bees, this is not always necessary, but a little smoke will ensure that you can move the bees safely and it will give you added confidence.

3 Pick up the complete nucleus box and move it to one side so that you can put the floor and brood box in its place, with the entrance facing the same way. At this stage, you don't want any frames in the brood box as you need room to insert those from the nucleus.

4 Wait for a minute to let the smoke take effect. Unstrap the nucleus box if you have not already done so, then remove the roof and take off the travelling screen. If you want, you can puff a little smoke over the top bars.

5 Use your hive tool to raise the left-hand lug of one of the outside frames, if you are right-handed, and then repeat with the right-hand lug. If you are left-handed, start on the right and move to the left. Grasp both lugs and lift the frame smoothly from the box. You may find it easier, particularly if your hive design takes frames with a short

lug, to grasp the top bar between the lugs rather than the lugs themselves.

6   The first frame can be awkward to remove but you can gain a little more manoeuvring room by sliding the flat end of the hive tool between the side bar and the hive wall and levering the frames away from the box. Then use the hive tool to break the propolis seal between this and the next frame. Try not to roll or crush the bees as you lift the frame. Bees will generally move out of the way if you give them a moment. They are pretty resilient and can survive a certain amount of rolling, but the less this happens the better.

7   Take this frame, with all the bees on it, and place it into the brood box. If you are using a National hive, you have the option to place the frames 'warm way' (parallel to the entrance) or 'cold way' (at right angles to the entrance). If you use warm way, put the frame from the nucleus at the front. Make sure that the side that was facing the nucleus hive wall is against the wall of the brood box. Building comb is not an exact science and the bees will have constructed adjacent combs so that they fit together. You want to keep this arrangement in the new box.

8   If you are placing the frames cold way, put the first frame against the equivalent sidewall in the brood box in relation to the entrance. Continue the same operation with the next and all the other frames until you have emptied the nucleus box.

9   You now have a gap that you need to fill – or the bees will do the job for you by building wild comb. Take the extra frames of foundation and, with a warm-way hive, place them at the back. In a cold-way hive, you can either put the frames to the side of the nucleus frames or you can move these into the centre of the box and fill up the gaps on each side. In the first arrangement, the brood nest will expand sideways and the bees will draw out the comb on each sheet of foundation as required. With the second, the bees will expand on each side and the comb may not be drawn out as uniformly. When the colony reaches its full size, the brood nest will be found centred behind the entrance.

10 Knock a corner of the nucleus box sharply on the ground to gather the bees still clinging inside together in a heap and then tip them on top of the frames in the brood box. You may have to strike the nucleus box sharply to dislodge the stragglers or you can brush them out with a bee brush or a handful of long grass.

11 When all the bees are in the new box, add the inner cover followed by the roof.

Key idea

When transferring the bees to their new hive, keep the combs in the same order.

Remember this

Make sure that the hive for your new bees is ready for when they arrive. Get help transferring them to their hive if you are not sure how to do this.

Catching a swarm

The other way of acquiring bees is by collecting a swarm. You could add your name to the list of swarm collectors held by your local association. However, although it is fun to do and it is nice to get something for 'free', sometimes swarms can land in tricky and even dangerous places. Be aware that not all swarms can be collected. Sometimes the risks are just too high or they prove impossible to reach. You will have no idea of the temper of the bees you collect, their propensity to swarm or their disease status, so your freebie might turn out not to be as good as you had hoped.

Collecting and dealing with swarms and the swarming process is covered in more detail in Chapters 9 and 10.

Try it now

✳ Keep only good-tempered bees that do not follow you when you leave the apiary.

✳ Obtain local bees that are in tune with the local conditions.

✳ Make sure that any nucleus you acquire conforms with (in the UK) the British Beekeepers' Association code of practice.

✳ Make sure that the box used for transporting your bees is ventilated.

✳ Prepare the colony's new hive and prepare its desired location before the bees arrive.

Focus points

✳ Acquiring bees is an exciting part of your beekeeping experience. Make sure that you chose the right bees for you and your situation.

✳ Before you purchase your new bees, it is very important to ensure that they do not have a notifiable disease.

✳ Transporting bees must be done with care, preferably in the evening or early morning when it is cooler.

✳ If possible, keep swarms away from other bees until both open and sealed brood can be checked for disease.

Next step

The next chapter describes how to inspect a colony, what to look for and how to handle the combs and the bees. It describes the steps you need to take as you inspect each comb and what you should expect to see in a healthy colony. The need for regular inspections is explained in relation to the swarming process. Advice is given about when to add more room in the form of supers.

8

Inspecting the colony

'Why bother?' some might say. 'Far too controlling.'

Knowing what is happening in a colony of bees means that a good beekeeper knows when to help and what to do. Not knowing the state of your colony reduces you to the ownership of a box of buzzing insects. Flies could fit the bill! There is more to bees and much more to beekeeping.

From the time your colony moves from its quiet winter state into spring activity, it is moving to the point when it may or may not want to swarm. Swarming is the natural act of colony division. Without some control, half of your colony may depart to give someone else your honey. I for one would like to have some control over that process.

When to inspect your colony

When should you consider opening a colony of bees?

First, the weather has an influence. What is cold and unpleasant for us makes bees more likely to be bad tempered – that is, more likely to sting. Weather that we find warm and comfortable is more likely to help the temper of your charges.

The time of day has an influence. Too early in the day is bad – later is much better. Any time, say, between the hours of 10.30 a.m. and 5.00 p.m. is fine, but don't think these times are set in stone. The quality of the weather is more important. Warm is good. Warm and sunny is better. The small-scale beekeeper can be much more precise than the busy commercial beekeeper.

Good-tempered bees handled well will cause no problems, but bees can become very annoyed by clumsy handling or just because they are 'touchy'. Disturbance of the hive may well provoke the harassment of the disturber, followed, maybe, by stinging. I'm afraid that bees don't discriminate between the particular person causing the problems and any other human close by. This willingness to deter any possible 'attack' on the colony can spread out to cover several houses and gardens down the road. Whatever is likely, it is up to you to ensure your neighbours' safety. Modern gardens can be very small. Consider scheduling your inspections away from weekends to mid-week daytime periods if possible. Your neighbours have the right to enjoy their property without being stung.

Remember this

When you handle the frames, be careful to avoid crushing bees or upsetting them. If a colony becomes very defensive, close up the hive and leave the inspection to another day. If you have very close neighbours, consider whether it is possible to inspect your bees during the week when they are not around.

The first inspection

There is no official date in the year for this. I think many beekeepers look too soon. Large urban areas are warmer than those in the country, so plants and bees will be ready earlier. Old-style suburbs provide excellent forage for bees, so the first inspection can be as early as late March, depending on where you live in the country. All instructions involving bees need to be viewed through a lens made from that rare commodity – common sense.

I am assuming you can light your smoker and keep it alight. You also need to feel confident, so wearing your protective equipment is a must. Light your smoker, then don your gear. Your hive tool has to live in your hand, held there when not in use by your third and little fingers. Proceed as follows.

Puff some smoke into the entrance. Make sure that some goes inside. Then give the bees a couple of minutes to react. The smoke causes them to engorge, drinking honey from open cells. This seems to calm them down a lot. Smoke can also be used to drive bees away from the frame lugs and top bars. Don't apply too much. Time and experience alone will teach you.

You must stand with the frames parallel to your body – at the side for cold-way hives and at the back for warm-way hives. Lift off the roof and place it upside down at your side nearby. Other things like supers or the inner cover can then be placed crossways on the roof, avoiding crushing bees where the parts touch.

Slip the broad, flat end of your hive tool between the edge of the inner cover and the box. People usually do this near a corner. You may have to push firmly if the bees have stuck the parts together with propolis. Lever the box and the inner cover apart with the hive tool, first at the corner, then down the opening gap at the side. Don't pull the hive parts apart. Break the adhesion of propolis with your hive tool. You will see propolis along the edges of the box and the rim of the inner cover.

If there are lots of bees on the underside of the inner cover, place the whole thing carefully upside down on the roof. The bees can be knocked or brushed back into the box but initially you may not feel confident enough to do this.

Next, look at the top bars. In a strong colony, you may well find bees covering (that is, walking on) all or some of the top bars. They may well be concentrated on just a few of the frames. To pull out frames at random annoys bees and teaches you nothing. You must remove a frame at the edge of the box. I invariably start at the side nearest to me. This is not vital. You can start at the opposite side if you wish, but you will be reaching over the bees to the frames you want to remove and some bees seem to dislike this. It's your choice.

Lift the first frame up gently to eye level. The aim is to move smoothly, not necessarily slowly, without jerks. Ask yourself some questions. Is it empty? How much food does it hold? A full brood frame will hold about 2.5 kg (5.5 lbs) of honey. Does it contain any pollen? Store the answers in your mind. All colonies need a minimum reserve of 5 kg (11 lbs) of food, the equivalent of two full frames.

The frame must be put somewhere. You can buy special rests that hang on the side of the hive where you can hook the frame. Hive stands can be built with room for the frame to be leant at the side. You may have a spare box to house the frame temporarily.

By removing a frame, you have created a space. Go to the next frame and lever it away from the rest of the frames into this space.

Continue to remove and inspect the frames. Hold each frame over the brood box while inspecting it. The queen may be on it and, if she falls off, you want her to fall into the hive, not on to the ground where she could get lost or stood on.

The real information is on or in the frames and the bees can get in the way, so brush them off back into the hive or hold the frame down in the gap and give it a sharp shake.

Once the comb is a bit clearer, you may well see liquid stores and pollen. The liquid honey will be shiny and the pollen will look solid in a variety of colours. A large slab of pollen may well indicate that the next frame will hold brood.

Remember this

Stand with the frames parallel to your body rather than reaching across the hive. Use smoke in moderation to move bees away from the frame lugs or an area of the comb you wish to inspect. Keep frames in the same order with the same faces together.

Checking the brood

Brood consists of eggs, larvae and pupae, or sealed brood. Of these, the hardest to see are eggs – and eggs are the part of the brood nest you must see when you inspect colonies. To see eggs means that you have a queen and you can safely remove things like queen cells, knowing the bees can rear replacements from the larvae that hatch from the eggs. Large larvae should look pearly white and be curled into a 'C' shape at the bottom of the cells. Very young larvae that have just hatched are about one-quarter to one-third of the diameter of the cell and lie in a whitish pool of brood food. Pupae lie in sealed cells, covered with a brownish capping. This can be very pale in new comb and it gets darker the more the comb is used for brood rearing.

As you start looking at the frames, the first one you find containing brood carries only a small patch. This may consist only of eggs. This is a good sign that the brood nest is expanding. The brood patch on the next frame will be bigger and it will be larger still on the next one. The size of the brood patch will increase on adjacent frames to a peak, and then decline. The brood nest is shaped something like a rugby ball if you think of it in 3D. It is surrounded by a shell of stored pollen, which is thickest at the back and above. It is thinner at the sides of the brood nest, with little or none below it. Quite a good amount lies between the brood nest and the entrance. Irrespective of the way the frames run, cold way or warm way, the brood nest will have the same orientation in relation to the entrance.

The cappings on sealed worker brood should be flat or slightly domed. Some bees produce a capping with a slight dimple in it and occasionally cappings can look slightly wrinkled. All this is quite normal and shows that the eggs you have seen are both fertile and fertilized. Drone cells are larger and the capping is domed. Drone brood may well be present in late March or early April, indicating that drones will emerge in about a fortnight and be capable of mating with virgin queen bees two weeks after that.

As you progress through the frames, the second comb removed is replaced in the gap you have created but pressed over to the side nearest to you. Thus the gap moves with you as you progress across the box. Don't lift out frames willy-nilly. This both upsets bees and makes it less likely that you will see the queen. You may also crush bees because, although combs are always parallel, they are not always straight. A bulge in one side of a comb is mirrored by a depression in the side opposite it on the adjacent frame.

Having looked at all the frames, you can push them back into their original places. With Hoffman self-spacing frames, you can do this in a block or two or three at a time. With plastic spacers, you may have to move each frame individually – another reason why I recommend Hoffman frames in the brood box. The final act is to replace the first frame you took out. It is important not to turn frames, and hence combs, around. Bees build a brood nest where brood patches exactly coincide and turning a frame through 180 degrees disrupts this pattern.

If your brood box is full of bees with, say, brood on eight or nine frames, then your colony needs more room. Smoke the bees down from the top bars, placing whatever type of queen excluder you have chosen on top, with the slots or bars at right angles to the frame orientation. You may have to scrape the top bars clean of wax or propolis so that the excluder lies flat. Collect the scrapings in a container. Do not just drop them on the ground. The super with its frames of drawn comb or foundation goes on next. Finally, replace the inner cover over the super, followed by the roof.

Key idea

Do record what you have seen – the date, the reaction of the bees, the estimated amount of food in the brood chamber, the weather, the number of combs of bees, whether there are drones present in the hive, and anything else you wish. Your records can be in any form that is meaningful to you. The important thing is that, when you look at your records, you know what happened at the last inspection and the state of your colony.

EXAMINING THE BROOD BOX

To really understand what is going on and monitor what progress the bees have made towards possible swarming, you need to examine the brood box on a regular basis and record your findings, at least during the swarming season. The interval between the examinations depends on the timings of the life stages of the brood, both queen and worker. Swarms generally issue when the first queen cell produced is sealed, assuming the weather is favourable.

Queen cells are built on the face of the comb, usually at the bottom or the edge. They are sealed nine days after the egg is laid. Therefore you need to do an inspection at least every nine days. This is awkward for most of us since our lives are built around a seven-day week. It does no harm to look regularly every seven days, once you see drones on the combs. A colony will not swarm before drones are produced. This regular examination is generally no longer needed once swarming has been successfully controlled.

Key idea

Inspecting the brood combs allows you to know the state of your colony and hence when to help and what to do. Handled carefully, you will be able to check the situation as you inspect the combs in turn. The brood is a very important part of the colony as its health dictates the strength of the colony and how it is likely to progress.

BROOD-AND-A-HALF

One way of reducing the disturbance of examinations is to use one brood box plus one super as the brood nest. This arrangement is termed 'brood-and-a-half'. Queen cells are built on the edges of comb or on areas where damage may have occurred. Bees seem to consider the bottom bars of a super placed on top of the brood box as such areas. Examination can then consist of (a) removing the supers down to the queen excluder and (b) drawing the 'half' slightly towards you and tilting it on the back edge so that you can inspect the bottom bars. A good proportion of the queen cells a swarming colony builds can be seen along those bottom bars. If there are no queen cells, replace the 'half' and restore the rest of the boxes to their original positions. Inspecting for swarming preparations them takes about ten minutes.

Remember this

Check colonies regularly for swarming preparations. Seeing eggs tells you that the queen was present at least three days ago. Although reassuring, you do not have to see her on every inspection.

ADDING SUPERS

Modern hives allow extra room to be added to a hive in stages. When the brood box is full of *bees*, add a super. When that super is full of *bees*, add another, and so on. Don't let the bees run short of room, as this is one of the factors that can trigger swarming. When you remove supers of honey for extraction, make sure that the total amount of space available to the bees remains the same and is not reduced in any way. This means that you need to have at least four supers available for every strong working hive.

Try it now

* Know what you expect to see in a colony before you take off the roof.
* Remove and replace frames smoothly and carefully so that you do not upset the bees. They may take it out on your neighbours.
* Know what healthy brood looks like so you can recognize when something isn't right and get help.
* Undertake regular examinations every seven to nine days to check for swarming preparations.
* Add another box (super) when the uppermost box is full of bees. Don't allow your bees to run short of space, as this can be a trigger for swarming.

Focus points

* Regular brood nest inspections are essential because they will tell you whether your bees are going to swarm and you can take steps to control this.
* The population of the hive will alert you to the need for more space in which the bees can store nectar. This is part of swarm prevention.
* The state of the brood nest will indicate whether the colony is expanding properly in the spring. If it is not, you need to check for diseases like nosema.

Next step

The next chapter describes the stages in the natural swarming process, from the production of play cells to their transformation into queen cells. The prime swarm leaves the hive and clusters. Scout bees identify a new cavity and the swarm flies to it and takes up residence. Casts, or afterswarms, are produced as virgin queens hatch in the original colony. Methods of collecting swarms are described.

9

Swarming

A colony of bees can be regarded as a 'superorganism' and swarming is thus its act of reproduction. In the spring, colonies expand. The queen lays eggs at an increasing rate, the brood develops and soon new bees are produced at a rate far exceeding the mortalities. The colony gets bigger and becomes overcrowded.

The queen produces a chemical scent, or pheromone, known as queen substance, that keeps the worker bees operating normally. Worker bees share food by trophallaxis, which results in pheromones also being circulated throughout the colony. As worker bee numbers rise, each gets a smaller portion of the same overall pheromone production. This stops them working normally and causes a good proportion of them to leave with the queen to form a new colony.

The stages of the swarming process

Within the colony, several stages can be observed. Your regular examinations will allow you to monitor the situation and take appropriate swarm control action.

DRONES

The first step in the swarming process is the production of drones. Weak and/or sick colonies don't produce drones. It is very easy to spot the larger drone cells, especially when these cells are capped over pupae. There is no exact date when colonies start producing drones. It can be delayed by poor weather, lack of food, altitude or low temperatures. Drones start to fly soon after they hatch. At about 14 days after emergence, they go to drone assembly areas for the purpose of mating with virgin queens who also fly there.

Key idea

Colonies that don't seem to be expanding in the spring are probably sick and may not swarm for that reason. The reason why they are not developing needs to be investigated and the necessary steps taken, but this will be covered later in Chapter 17 where we talk about diseases.

QUEEN CUPS

The next sign that the hive is getting ready to swarm is the production of queen cups, or 'play cells'. These are the bases of queen cells (Fig. 9.1). They look rather like acorn cups. Many are built on the edges of populated combs, the bottom edges of combs in an upper box (where you are using a brood-and-a-half or two brood boxes – double brood – for brood rearing), and in places where damage or holes might appear. When these queen cups are occupied by eggs or larvae, they become queen cells and the workers begin to draw out the cell walls.

At that point, it is wise to assume that the colony is preparing to swarm. Discovering this point is the purpose of your regular examinations, which, in the UK, start in late March or early April, depending on where you are in the country, and may have to continue until late June or early July. Again, this date may be later in cooler or higher regions.

Figure 9.1 A sealed queen cell

Once a colony has started swarming preparations, a successful application of swarm control methods usually means that such regular inspections are no longer required. If nothing is done and the swarming process is allowed to proceed naturally, the steps are as follows:

1　Eggs are laid in batches. The queen cells in a colony are never all at the same stage of development.

2　The first egg laid in a queen cup will be the first to hatch, and so on. All eggs hatch after three days.

3   The larva is fed royal jelly lavishly and eats lavishly. This jelly looks a bit like condensed milk although it doesn't taste like it. By the time the larva pupates, six days later, it is lying in a bath of this rich food.

4   The queen cell has now been extended by the workers to about 3–4 cm (1–1½ in.) long. It is tapered and has a rough, textured surface. Queen cells hang vertically from the face of the comb with the open end at the bottom. The larva seems to retain its position in the cell by being 'stuck' in the excess brood food.

5   On the ninth day after the egg hatched, the lower end of the cell is sealed over with wax. It is on this day or the next when the swarm usually issues from the colony.

6   About half the bees, mainly the young or middle-aged ones that are one to three weeks old, form the swarm, leaving the colony together with the mated queen. Many of the older bees, the foragers, stay with the original colony.

THE CLUSTER

The queen substance pheromone is very attractive to worker bees. In a swarming situation, one component in particular attracts the workers to the queen and another keeps the swarm cluster stable. When the swarm leaves the hive, the queen usually settles on some convenient place near the hive, say 3–10 m (20–30 ft) away. She can pick some bizarre places. Some I have found include: a curtain pelmet, a hollow cast-iron lion, under a car mudguard, down a street drain and inside a compost bin. The swarm bees join her and form a cluster.

While clustering in this place, the bees search for a suitable home for the colony. It can be located anywhere within a radius of around 300 m (1,000 ft). I have seen bees investigating empty brood boxes at my home about a week before the swarm arrives. I am convinced that, as far as this first, or prime, swarm is concerned, the scouts start to look for new homes while the queen cells are being built. Scout bees communicate using dances that are similar to those which tell other foragers of nectar sources. They will inspect one other's finds until a decision is made and the whole swarm departs.

Cold weather with rain will make the cluster hang in the original location for longer than usual. If it is in the shade, this can have a similar effect, while direct sun shining on it makes the bees unsettled.

FLYING TO THE NEW NEST

Most of the bees in the swarm don't know where to go because only the scouts have been out looking. These scouts fly forward through the swarm in the direction it needs to fly, then rise or drop, fall back behind the mass of bees before flying through the swarm again. The queen, of course, flies with the swarm and her pheromones keep the bees together. Some of the scouts fly to their destination and then fan at the entrance of the chosen cavity. They produce another pheromone from the Nasonov gland at the tip of the abdomen, which attracts the swarm to the desired place. They stand facing towards the cavity, expose their Nasenov glands and fan their wings vigorously to distribute the pheromone.

The swarm enters the cavity and clusters inside at the top and towards the back. About half the weight of a swarm consists of the nectar the bees carry in their crops. This is used, not just for food, but to provide the fuel for building comb. Wax-making clusters are established and small particles of wax deposited on the upper surface of the cavity. Lines of wax start to appear on which the comb is then built. Some of the honey in the bees' crops is deposited in these cells. As soon as possible, the queen starts to lay in the initial cells. The swarm aims to collect enough nectar for winter and to produce a strong enough population to ensure the colony survives.

Key idea

Swarming is a natural process that follows definite timings related to the development of the new queen(s). Swarms are kept together in flight by the queen's pheromones. The Nasonov pheromone produced by workers helps guide the swarm to the new nest.

For the first three weeks, the swarm colony will be losing bees. Older bees will be dying but the new brood will still be developing. It is not until the first new bees emerge that the swarm can begin to replace its losses.

Virgin queens

Back at the original colony, after the departure of the swarm, all goes quiet. A number of queen cells contain new queens at different stages of development. The colony contains a lot of worker brood and some drone brood. In the week following the swarm's departure, this brood will be hatching at the rate the original queen laid the eggs. There are likely to be 1,000 bees hatching from the combs each day, to some extent replacing the bees lost with the swarm.

The first queen cell started, developed and sealed will produce its virgin queen about seven days after the prime swarm departs. During this week, bees pay a lot of attention to the queen cells. They are obviously aware of how old they are because wax is removed from the tip of the most advanced as the queen inside nears emergence. Cells where the brown parchment-like cocoon spun by the occupant is exposed are known as 'ripe' queen cells.

Casts or afterswarms

When she is ready to emerge, a virgin queen starts to bite a circular track through this tip, just back from the extreme end. The first virgin emerges and joins the colony. That day or the next, she may leave with a smaller swarm or cast, also known as an afterswarm.

This second swarm clusters and then flies away, just as the prime swarm did. Back in the hive, two or three cells may be ready to hatch. One virgin may well be loose on the frames. Her instinct is to 'pipe'. This produces a note around middle C.

Peep, peep … peep-peep-peep is the type of sound we humans hear. It can be heard outside the hive. The virgin is looking for occupied queen cells in order to kill the occupants. The other virgins may well pipe from within their cells in reply to the free-running queen but this sound is more muffled.

A second cast may issue a day or so after the first. There may even be a third or fourth cast. Some of these later casts can be very small. All will frequently contain more than one queen.

Eventually, enough bees and queens have left the original colony. The bees allow any remaining free-running queens to fight until only one remains alive. The occupants of any queen cells still occupied are stung to death by this queen and the workers remove the bodies. The surviving queen flies from the hive to mate, returns and becomes the new queen of the original colony with all its food reserves. The original colony may now have divided itself into five or six new units. Of these divisions, the prime swarm, possibly the first cast and the final remnants of the original colony with its new queen have the best chance of natural survival. About 75 per cent of natural swarms do not survive their first winter without help.

Key ideas

* The swarming process follows a defined procedure linked to the development cycle of the queen. Swarming preparations do not begin until drones are present in the colony.
* When an egg is laid in a queen cup, this becomes a queen cell, the walls are elongated by the workers and the occupying larva is fed copious amounts of royal jelly.
* The swarm leaves the hive and clusters nearby until a consensus is reached on the location of the new nest. The swarm flies there and takes up residence.
* Virgin queens hatching in the original swarming colony may lead out further swarms (casts). Eventually, one virgin queen kills any other free-running queens and the workers destroy any remaining occupied cells. She flies out to mate and returns to head the colony.

Collecting a swarm

Help to survive can come in the form of beekeeper assistance. The most obvious help is for the beekeeper to collect the swarm, put it in a hive and possibly feed it. So how do you collect a swarm?

The optimum time to collect a swarm is while it is hanging in a cluster and the scouts are deciding on a suitable place to live. Once it has taken up residence in a cavity (a chimney or cavity wall, for example), it is much more difficult to remove and you might need to make holes in masonry or brickwork, which is not something to undertake lightly or without the owner's permission. In such a case, I advise you to consult experienced beekeepers from your local association.

To collect a swarm, you will need:

▶ a container in which to put the bees

▶ a cover for the container through which the bees can breathe

▶ a mist sprayer full of clean water

▶ your veil, smoker, smoker fuel, matches and hive tool

▶ a long piece of string.

You may also need secateurs or loppers depending on the cluster's location.

The emergency receptacle can be something like a strong cardboard box. The cover should be a material with a fairly open weave, like plain net curtains or a sheet. It needs to completely cover the opening in the receptacle with enough surplus to hang over the sides and be tied on with the string.

Key idea

The best receptacle for swarm collection is a straw skep. This is strong and light and has a rough surface to which the bees can cling easily.

When bees form a cluster, some cling to the surface, be it a wall, a branch or a bush, and the rest cling on to them. They tend to face upwards with their bodies slightly overlapping, like tiles, to shed any rain. In the classic swarm, the cluster is free-hanging. However, if it is in a bush, for instance, you need the owner's permission to cut branches away until you can put the receptacle right under the swarm. For safety's sake, you need to be on your own feet or a very steady set of steps when collecting a swarm. You also need to put on your veil. Swarms are generally good tempered but occasionally one isn't. You must be prepared.

1 Place your covering cloth nearby. Lightly wet the surface of the swarm cluster with the mist sprayer to make the bees cling together more tightly. They will then tend to fall in a clump when shaken.

2 Hold your box or skep beneath the swarm. Lift it until the swarm is inside it as far as possible. Knock or shake the bees firmly into the box. The more the box surrounds the swarm, the less chance there is for bees to fall outside it. The aim is to get the queen, with her stabilizing pheromones, inside. This is more important than getting all the bees in.

3 Climb down to the ground carefully if necessary and then cover the box with the cloth. Invert it and place it, opening downwards, as close to the clustering point as possible.

4 Spread out the cloth and prop up one side of the box to make an entrance. There will be many flying bees in the air and the aim is to collect as many of these as possible.

A good sign that you have collected the queen is when a large number of bees start to fan at the entrance and expose their Nasonov glands. The pheromone they produce should attract the flying bees to the box. The original clustering site will still be attractive to the flying bees. Bees produce a footprint pheromone, which they will have left there. It is quite attractive to bees. This, plus the traces of queen pheromone, attracts bees back to this location. Use your smoker to smoke bees away from the site. Shaking branches to make any returning bees take flight is also a good idea. Once aloft, the bees are more likely to join those down on the ground.

Eventually, it will become obvious that the flying bees are joining those in the box. You have given the bees what they were looking for – a suitable home. It is not the one they were aiming for and they may still decide to leave your box and fly to their chosen cavity. When you feel you have as many bees as possible in the box, gather up the edges of the cloth around it and tie them in place with the string, making sure that there are no gaps from which the bees can escape. Invert the box so that the bees can breathe through the cloth. An alternative but not always practical strategy is to return at dusk when all the flying bees will be in the box and take it away then. If there are still some bees flying or clustering on the original site, they will return to their original hive when they realize the swarm has gone.

Not all swarms hang neatly like this. Some may perch on a fence post with wire netting running through the cluster. In this case, the bees can be brushed down into the container using a goose flight feather or a bee brush. If neither of those is available, a handful of longish grass will do the job. If the bees are in a dense bush or hedge, contrive to place the box with one edge touching the swarm, then use the smoker to drive the bees gently into it. Sometimes you can balance the box on top of the hedge and encourage the bees to climb upwards into it. Once you have at least half the bees in the box, place it on the cloth, spread out on the ground nearby, give it an entrance and smoke the remaining bees into the air. The process then proceeds as before.

Remember this

When collecting a swarm, the key is to get the queen into your collecting box. The other bees will then join her.

Key idea

Collecting swarms can be regarded as a public service, especially if they have clustered in a public place. Use this opportunity to talk to bystanders about bees.

Hiving the swarm

Bees remember where their hive entrance is as a point in space. Any significant displacement of this will cause disorientation at the entrance. When bees swarm, they choose to forget their old nest unless there is an immediate loss of their queen. It is possible to take the swarm you have collected and put it in a hive anywhere you like in the apiary, even just a metre from its original home. If your swarm is a 'gift from nature' and of unknown origin, you can certainly hive it wherever you like. There is around a one-per-cent possibility, that such a swarm might carry American foul brood and a greater chance of European foul brood (see Chapter 17). It is a good idea to hive these 'wild' swarms on foundation and to reduce the chance of their bees drifting into other colonies in the apiary.

Bees in a swarm are ready to produce wax and build comb. They do it very well. It is sensible, therefore, to hive a swarm on frames of foundation. You may not have enough such frames ready when you get your swarm, but you do have a little time to make them up.

Go to the place where you wish to hive your swarm. There are basically two ways of doing this: the dramatic, or traditional, way and the less fussy way. I suggest you use the traditional way once because it is great fun and then the other way next. You can choose which you like best.

Before you start, gather together the equipment you need. This is a floor, a brood box full of frames of comb and/or foundation, an inner cover and a roof. You will need your veil, smoker and hive tool.

TRADITIONAL HIVING

This operation should be carried out after around 7 p.m. so that the bees will go into the hive and stay there rather than deciding to fly off to another cavity. You will need a flat surface the width of the hive to slope up from the ground to the hive entrance.

1   Place your swarm in its container, opening downwards, on the slope.

2 Undo the string and spread the cloth out to hang over the sides of the slope. Smooth it out as much as possible.

3 Raise the container about 300 cm (12 in.) and give one firm jerk downwards, stopping short of hitting the slope. Most of the bees should end up on the slope. They will start to crawl up towards the hive.

4 Brush some bees gently towards the entrance if they are a little slow. They will start to enter the hive. They may block the entrance and crawl up the front, so try to stop this by brushing them back down.

5 You can also enlarge the entrance by inserting small blocks at each side. You can gently guide the stragglers inside by the judicious use of a little smoke.

It can be fascinating to watch the bees marching up to the entrance and some people have been lucky enough to see the queen among them. Once the majority of the bees are inside, you can remove the slope and the cloth and the blocks in the entrance if you have used them. There is no real need to feed the swarm immediately but food in the form of fondant or syrup can be given after a day or so, particularly if the weather is bad.

DIRECT HIVING

This hiving method is much simpler and quicker. As well as the equipment noted previously, you will need an empty super or an eke. This is made from four pieces of wood, about 8 cm (3 in.) deep and the lengths of the hive sides, nailed together at the four corners. It has the same cross section as the hive and forms a temporary extension to it.

1 Lift the brood box, its frames, inner cover and roof and put them to one side.

2 Place the empty super or eke on the floor. Throw the swarm into this.

3 Place the brood box full of frames on top and add the inner cover and roof.

4 Some time the next day, take off the roof and place it nearby upside down. Lift the brood box and inner cover gently on to it. Remove the super/eke and replace the brood box, etc., on the floor.

Later swarms in June or July can be casts, each containing several virgin queens. They are very prone to stay the night and then leave the next day. Place a queen excluder on the floor before adding the super/eke. A wire excluder works well here. Hive the swarm, shaking it onto the excluder and completing the process as described above. The workers can climb down through the excluder and fly from the hive, but the queen(s) cannot. Forty-eight hours later, remove the excluder as well as the super/eke and cut off any comb that may have been built from the bottom bars of the frames in the brood box. By this time, any fights between the virgin queens will be over and the swarm is very likely to stay put.

CHECKING FOR DISEASE

Swarms may carry disease. Two, American foul brood (AFB) and European foul brood (EFB), are notifiable. This means that, if you even suspect your bees may have either, you are required *by law* to inform the authorities. Details are given in Chapter 17. EFB shows its symptoms in unsealed brood which, in the case of a prime swarm, will be present about seven days after hiving. Signs of AFB show up in sealed brood and will be present 9–21 days later. For a cast, there will be a longer delay because the virgin queen must first be mated. Leave a cast for 10–14 days before having a look. Inspecting it sooner could result in the virgin getting lost if she is out on her mating flight when you open the hive.

If you have any doubt at all about the disease status of your swarm (or colonies for that matter), contact your seasonal bee inspector, who will be only too pleased to come out and help you.

Remember this

Hive a swarm away from your apiary if possible until you can check both open and sealed brood for disease.

Bait hives

I would suggest that every apiary should include at least one bait hive, especially in built-up areas. This is a hive set out specifically to attract a swarm to take up residence. It certainly offers the easiest way to collect a swarm!

Scout bees are looking for a suitable nest site for the swarm. What better place than a beehive with frames and, preferably, at least one drawn comb and foundation? The box will smell of bees, which indicates that it is a suitable home. The comb inside will mean that the bees can take up residence and the queen can begin laying virtually immediately.

The best position for your bait hive is the one where the bees will naturally be looking for a cavity, 3–4 m (10–13 ft) above the ground. This is probably not a practical location for most urban apiary sites, but roof-top apiaries should have an advantage here. In a garden it's not vital, but just try to raise the bait hive as high as is practicable to increase the chances of occupation.

You can purchase swarm lures to place inside the box to add to its attractiveness. These contain the chemical produced from the Nasonov gland which bees use to attract others to their new home. Lemon balm produces similar chemicals and some beekeepers have had success by rubbing its leaves around the inside of the bait hive.

Keep an eye on your bait hive during the swarming season. You may well get an early indication of a swarm's arrival when you see scout bees fussing around the box. If the consensus is that this is the best home, the swarm should arrive in a few days. Mind you, if other scouts have found a more attractive cavity, you may find that the interest in your hive wanes suddenly and dramatically. This can also happen if another beekeeper catches 'your' swarm while it is clustering and takes it back to his/her own apiary.

Remember this

Ideally, place bait hives 3–4 m (10–13 ft) above the ground. In practice, locate them as high as is feasible.

Try it now

* Collecting a swarm can be fun but make sure that you do it safely.
* Hive a swarm over a queen excluder to stop it absconding.
* Check swarms for disease.
* Put at least one bait hive in your apiary.

Focus points

* A colony is preparing to swarm when the queen lays an egg in a queen cup, turning it into a queen cell.
* When the first queen cell is sealed, the original queen swarms from the colony together with half the bees.
* A week later, virgin queens begin to emerge and can leave in a cast with half the remaining bees.
* Eventually, one virgin will remain, and she flies out to mate.
* A swarm is easiest to collect when it is clustering. If it is in an awkward place, do not risk your safety trying to collect it.
* A swarm can be hived anywhere in the apiary.
* A bait hive can attract a swarm, which cannot then be a nuisance elsewhere.

Next step

The next chapter discusses swarm prevention and swarm control, particularly in relation to the urban setting. The former can be achieved by giving colonies more space. The latter requires intervention by the beekeeper. A colony can be regarded as consisting of the queen, the brood and the flying bees. Swarm control is based on the separation of one of the colony's parts from the other two.

10

Swarm prevention and control

Swarm prevention and control are a vital part of urban beekeeping.

As we have seen, the average size of cavity chosen by swarms is 40 litres (11 US gallons). This allows average-sized swarms to reach a size that fills that cavity by May or June the following year. The congestion (crowding) that results from this is one of the prompts that starts the swarming process.

Beekeepers often talk about swarm prevention and control as though these terms were interchangeable, but this is misleading. Swarm prevention refers to activities that stop a colony from starting swarming preparations. Swarm control refers to steps taken to control or modify the natural course of events in the swarming process.

Swarm prevention

The provision of unlimited room would seem to be the answer for swarm prevention but colonies in extra large cavities still swarm. Some research at Rothamsted Research Station in Hertfordshire, UK, showed that, even given plenty of room, 27 per cent of the colonies under test produced swarms. The feeling of many beekeepers is that most bees try to swarm. Several of the colonies in the Rothamsted test started queen cells but aborted the process of swarming. There is no doubt in the experience of many practical beekeepers that all colonies would swarm if no extra space were given to them at all.

What is enough room for an overwintered colony? I am talking here of British Standard boxes. With beehives, additional space can be added or removed as required. In my opinion, a British Standard brood box or a brood box plus a super is perfectly adequate for most colonies in the British Isles. The excessively prolific bees obtainable today may need more room. I am afraid I know nothing of them. In my opinion, they are less than sensible for our climate.

In addition, apiaries with small numbers of hives, say one to ten, require four supers per hive. This allows you to exchange empty supers for full ones so that at no time does a colony have its internal volume reduced. Sudden strong congestion caused by the removal of supers without their replacement with empty ones is very likely to result in the start of queen cell building.

GIVING EXTRA ROOM

The main method of swarm prevention is to give the colony additional room. If bees are forced to store nectar in the brood nest, this will again give them a feeling of congestion. Room – that is, more supers – should be added before it reaches this stage. The problems caused by giving supers too soon are vanishingly small when compared with those caused by supering too late. I have known a super containing drawn comb be filled in three days. Hence the golden advice: 'Super ahead of requirements.'

Some bees are more prone to swarm than others. Bees of the Carniolan type are very good tempered but very prone to swarm because they expand very rapidly in the spring. Bees that have a lower tendency to swarm, or are more easily controlled when they do, are less of a problem in the urban setting. Such bees are not easy to acquire and take some effort to maintain. However, they are certainly worth it.

Remember this

Giving a colony extra room when it is needed is the key to swarm prevention.

Swarm control methods

Outside the swarming season, you need to decide on a method to use and acquire or make any extra equipment needed to put it into operation. If you have more than one colony, you must realize that your bees are not going to attempt to swarm in turn but are likely to do it all at the same time! Each hive must have its own set of spare equipment.

Many years ago, around 1990, I reviewed the methods of swarm control advocated by many writers on beekeeping. I felt they were all dealing with the same natural process and should have a lot in common. I decided the basis was to think of the colony as if it consisted of three parts: the queen, the brood and the flying bees. Separating one of those parts from the other two formed the basis of all good methods. In addition, most methods require intervention at two places in the natural swarming timetable. These are:

▶ early in the building of queen cells before any are sealed

▶ just before any of the young queens emerge from the queen cells.

MARKING QUEENS

Some swarm control methods require the beekeeper to find the queen. This is a lot easier if she is marked with a spot of paint on her thorax. By following an international colour code, it is possible to identify the year in which the queen was raised. Some beekeepers find one or two colours, such as yellow and white, much easier to see than the others, blue, red and green. You can use whichever colour you wish but, if you don't follow the code, you will need to keep a record of when your queen was hatched.

The colours are paired with the last numeral of the year. This sequence is repeated twice in a ten-year period. Since most queens rarely live beyond the age of five, this system both marks and dates the queen.

The agreed sequence of colours is white, yellow, red, green and blue. This table makes it easier to follow:

Year ending	Colour
1 or 6	white
2 or 7	yellow
3 or 8	red
4 or 9	green
5 or 0	blue

If you purchase a queen, you can ask that she be already marked. If not, it is not too difficult to do yourself. Suitable marking paint and aids to do the job can be bought from equipment suppliers. One simple piece of equipment is a Baldock queen cage. It is also called a 'crown of thorns' cage. It consists of a ring of composite material with spikes round one edge approximately 3 mm ($1/8$ in.) apart. Threads are stretched across the ring to form a grid. When you locate the queen, place the cage over her on the comb and press it down gently to hold her still. Adjust the final position of the ring so that the queen's thorax is visible in a square of the crossed threads. You can then apply a spot of paint with something like a matchstick. Some paints have an inbuilt brush but I generally find these are too broad and smother the queen in paint rather than just depositing a small, neat blob. Practise on drones first. You don't want to reduce your queen to cubes in your first attempt or coat her with too much paint.

Remember this

There are numerous methods of swarm control, all based on the same principle. The steps in swarm control are linked to the biological development of the new queen(s) and interventions cannot be delayed because of bad weather or other commitments.

THE NUCLEUS METHOD OF SWARM CONTROL

One of the simplest methods of swarm control is the 'nucleus' method. This involves separating the queen from the other two parts of the swarming colony, the brood and the flying bees. It can be done rather crudely by just removing the queen. Since she cannot survive on her own, she can be killed with a quick pinch. However, since things can go wrong and the queen would be the means of putting them right, it is much safer and kinder to keep her in a nucleus. Having said that, if all your hives are full of bees and you have no spare equipment, then killing the queen may be the only option. Be prepared!

The nucleus consists of a small number of combs in a box. Nucleus boxes are built to hold a few frames, usually five. If you don't have one, then you can use a spare brood box. You need to have sufficient frames of comb or foundation to fill either box. If you don't have enough frames to fill the brood box, use a dummy board (a wooden form the same size as a frame) to contain the bees on the available frames.

The following method assumes that you are using a five-frame nucleus box. Let us also assume that in your regular examinations you find queen cells and that none is sealed.

1 Remove the frames from the nucleus box. Find the queen in the swarming colony. Put the frame she is on, together with the adhering bees, into the nucleus.

2 Take two combs containing a good amount of food and place them on each side of the frame with the queen. The two outside frames in the swarming colony will probably contain the most food. Push all three frames to one side of the box.

3 Brush all the bees from two more brood frames into the nucleus. Fill up the box with two of the frames that were originally in the nucleus and add the inner cover and the roof. Move the nucleus approximately 1 m (3–4 ft) from the swarming colony.

4 Check the remaining frames in the original colony and note any queen cells that contain larvae. Mark the position of each with a drawing pin in the top bar directly above them. Push all the frames over to one side (cold way) or the front (warm way) and fill the gap with the spare frames originally in the nucleus box. Restore any supers and close up the hive. The bees cannot swarm without a queen. They have several queen cells, which cannot emerge for at least one week.

5 The nucleus with the queen will lose all the bees that are used to flying from the original colony. When they go out foraging, they will return to the original site and the original colony. The additional bees brushed from the two extra combs help to maintain numbers as they are largely nurse bees and will help to feed the larvae. The food in the combs should keep the bees happy for a while and the nucleus colony should start to build up as the queen lays a new brood nest. If the weather is poor or little food is available, after 24–48 hours feed the nucleus with fondant.

6 One week after this manipulation, go into the original colony. Check that the queen cells you marked are still there. Do *not* shake the combs carrying queen cells or you may damage the developing queens. Choose just *one* of the cells you marked that contained a larva. Brush the bees off that comb to make sure that there is only one queen cell present. Break off any others. Inspect the remaining combs, shaking the bees off each frame and destroying any queen cells – or anything you think looks like a queen cell. Just pull them off. When you are sure that only one queen cell is left, close up the hive. Do not look again for three weeks.

7 If the new virgin queen is out on her mating flight and returns when you have the colony open, she is likely not to

enter the hive and you will be left with a colony that has no means of rearing a replacement.

8 Worker larvae can be turned into queens of sorts up to the point they are three days old. One week after you removed the queen to the nucleus, any unsealed brood left from eggs laid on that day will be too old to be reared as new queens. Leaving just one queen cell with no others should ensure that, when that hatches, the resulting queen will go on to head the colony. Having one young queen needing to mate on which the future of the colony depends is normal in our bees. Leaving two cells won't help as either one queen will kill the other, leaving the same situation, or one will fly off with a cast, again leaving one on its own.

9 Check the brood nest after three weeks. Look for eggs. If you see none, put in one brood comb containing eggs and very young larvae from the nucleus, swapping it with a comb from the colony. However, shake the bees off each before you do the transfer, to make sure that you don't transfer the queens as well (you may have a queen in the colony). If the colony is queenless, queen cells will be started on some of the introduced larvae over the next week. The odds are that the new queen in the original colony hadn't started laying and one week more is all that was needed. The choice for later on is whether to keep two colonies or unite them back into one.

THE ARTIFICIAL SWARM METHOD

This method requires a brood box plus combs, a floor, inner cover and roof (Fig. 10.1). When you find queen cells in your regular examinations, proceed as follows:

1 Move the original brood box temporarily on to the new floor placed nearby.

2 Put the new brood box in its place and remove two combs near the front (warm way) or from the centre (cold way).

3 Find the queen and transfer her on the comb she is on, together with the adhering bees, into the space you have made in the

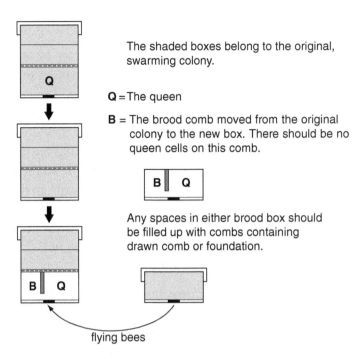

The shaded boxes belong to the original, swarming colony.

Q = The queen

B = The brood comb moved from the original colony to the new box. There should be no queen cells on this comb.

Any spaces in either brood box should be filled up with combs containing drawn comb or foundation.

flying bees

Figure 10.1 The artificial swarm method when the queen is found

new box. Remove any traces of queen cells from that comb. Replace the second comb you removed to complete the set.

4　On the original site, you now have the queen and almost no brood. You have separated the queen from the brood and the flying bees.

5　Cover this box with the queen excluder and add the supers from the original colony, the inner cover and the roof.

6　Push the frames together in the original brood box on the new floor and use the displaced comb(s) from the new brood box to fill any gaps. This box is queenless.

7　Replace the inner cover and the roof. Move the hive to a new position about a metre away from the original colony. Face the entrance to one side, at roughly a right angle to the entrance of the hive containing the queen.

One week later, do the following:

1 Examine both boxes in turn. The one containing the queen should show evidence of egg laying and comb building. Several queen cells will have been built in the queenless box. Reduced these to one. Do not leave two. This risks the first virgin to emerge leaving the hive with a massive cast containing the bees that would have been in the prime swarm, plus those of the first afterswarm. As soon as your new queen is laying, you can start to transform her colony into your main honey producer.

2 Over a couple of days, rotate the hive to face the same direction as the original colony. Swap as many combs of brood as can fit in – minus bees – with broodless combs from the young queen's hive. Move the new queen's hive closer to the original colony. On a day when the bees are working hard, transfer the supers from the original colony to the new queen's hive as follows:

   a Put an excluder over the new queen's brood box.

   b Place a new super with frames, preferably containing drawn comb, on top.

   c Spread a single sheet of newspaper over the hive. Pin it to the box if necessary to keep it in place.

   d Place the supers from the original colony on top of the paper.

   e Move the original colony at least 1 m (3 ft) away or turn it through 90 degrees to face a different direction.

   f Add the inner cover and roof to each hive.

   g The new queen will now have most of the brood, the flying bees from both colonies, and the colony is very unlikely to make swarm preparations that season.

   h Make sure that this colony does not run short of space.

IF YOU CANNOT FIND THE QUEEN

Maybe you cannot find your queen. Maybe you want to keep the manipulations as brief as possible in your urban situation. What do you do then?

Proceed as though you had found her:

1 Move the old box on to the new floor. The queen is inside.

2 Transfer one comb containing brood in all stages, particularly eggs and young larvae, plus its adhering bees, into the new brood box.

3 Put the excluder on top, followed by the supers, inner cover and roof.

4 Move the original brood box to one side, as before. Close up the frames and fill any gap with additional frames. Add the inner cover and roof.

One week later, check the following:

▶ If you find lots of queen cells on the single comb you transferred into the new brood box, the odds are that the queen was not on this comb.

▶ If there are no queen cells but eggs laid in new cells, you did transfer the queen. This is equivalent to the first stages of the artificial swarm and you can proceed as above.

▶ If there are queen cells on this comb, I suggest that you reduce them to one. They have been built where the eggs were on the transferred comb, so the resulting queen will be properly nourished during her development.

You now know that the bees in the other box are 'queenright', that is, they have a queen. There will be no queen cells because moving the box from its original position has diverted the flying bees back to the original site. With no flying bees, the colony will not swarm.

Again, when the new queen is laying, concentrate the brood, flying bees and supers with the new young queen, as described previously.

Key idea

Make sure that you have all the necessary equipment for your chosen swarm control method for each of your colonies. They are likely to swarm around the same time.

CLIPPING THE QUEEN

Some beekeepers clip one of the queen's wings so that she is unbalanced and cannot fly. Others regard this as mutilation. Whatever your view, simply clipping the queen does not prevent swarming. When she tries to leave the hive, she falls to the ground and gets lost, although sometimes a group of bees will cluster around her. The majority of the swarm returns to the hive. In the hive, the virgin queens continue to develop in their cells. When the first virgin queen hatches, the colony is still in swarming mode and she will leave, together with not only the original bees that tried to leave with the old queen but also the bees that would have left with the first cast.

To prevent this happening, you need to recognize that your original queen has in fact gone. There will be no eggs in the cells and the bees may have started to fill empty brood cells with nectar. At this stage you need to check all the combs for queen cells. Just note them; do not destroy them. The first one you see may be the only one in the hive. Assuming there are several, choose one that is a good size and has a rough exterior. Then destroy all the rest. The resulting virgin queen will then be accepted as the colony's new queen. She will mate and return to the hive to head the colony.

Remember this

Clipping a queen does not prevent swarming but simply delays it until the first virgin hatches, unless you intervene. You do not need to find the queen in order to control swarming.

Key idea

Don't let your bees take you by surprise. Begin regular examinations before they start swarming preparations. If you have a plan in place and the necessary equipment, there is no need to panic. Just follow your plan.

The importance of swarm prevention and control

Stopping your bees from swarming is vital in an urban environment. Some will tell you that you should allow this natural process to occur and do nothing to prevent it. That may be their view but I do not believe you can do this when you are surrounded by non-beekeeping neighbours.

To most people, bees are frightening – if not terrifying – insects. OK, we know this is not the case but you have a responsibility to ensure that your bees do not cause a nuisance to your neighbours. This includes doing your utmost to prevent them from leaving their hive in great numbers and clustering in a tree or somewhere else in the next-door garden. It also includes doing everything you can to make sure that the bees do not take up residence in a cavity wall or chimney in your neighbour's house. Once a swarm has entered such a cavity, it is extremely difficult to extract it safely. Sadly, the best way to deal with this situation is to kill the bees and then block up or cover any entrance gaps with a bee-proof mesh. You have decided to keep bees and take care of them. Letting them swarm and end up in a situation where they are killed is not good husbandry.

Beekeepers can get very uptight and anxious about swarm control. *Swarm control is not difficult.* However, you do need to understand it and be disciplined in carrying out the various stages at the appropriate times.

The best thing to do, well before the beekeeping season begins, is to sit down somewhere quiet and think about what happens. Think about the three parts of the colony and understand why separating one from the other two means that the colony cannot swarm. To start with, I strongly recommend you follow one of the methods outlined above. Follow it closely. Don't think you can leave the second intervention, say, for ten days rather than a week. By then, the queen cells will have hatched and you will have lost at least one afterswarm and half your bees. From a purely selfish point of view, reducing the colony population by half means that you are reducing the potential honey crop by the same amount because you will have lost a large proportion of the foraging bees. Even if you don't want to remove the honey, you have reduced the colony's ability to collect sufficient food to sustain it over winter.

Key idea

When you find queen cells, don't panic. Go and have a cup of tea and work out what you are going to do.

If at all possible, study swarm control *before* your bees are likely to start their swarming preparations. Decide on the one method you are going to use. Learn it thoroughly and understand what you are doing at each stage. If it helps, write out the stages on a card and take this to the apiary. Keep notes of what you do when. Acquire the necessary additional equipment needed and make sure you have enough for each of your colonies. They are very likely to decide to swarm around the same time so you won't be able to use a spare box for one colony and then another, as the first one will still be using it.

Remember this

Seeing a swarm in flight, or collecting it, is exciting – for beekeepers. For the general public, it can be terrifying. Make sure that any swarm you see isn't from your bees!

Try it now

* Swarm control is not difficult. Take time to study and understand the principles.
* Learn to recognize when your colony is preparing to swarm.
* Realize that this is something that needs to be prevented or controlled.
* Take the necessary steps at the appropriate time.
* Choose and learn a method you want to use. Write it down and take it to the apiary if that helps.
* Understand the importance of swarm control in an urban environment and the responsibility you have taken on by keeping bees there.

Focus points

* Successful swarm prevention means the colony does not attempt to swarm.
* Swarm control means you prevent a swarm from leaving the hive.
* Swarm prevention and control are a vital part of urban beekeeping.
* Marking a queen makes her easier to find as part of swarm control measures.
* Clipping a queen's wing only delays the swarming process. It does not control it.
* Regarding the colony as consisting of three parts, queen, brood and flying bees, all swarm control methods are based on separating one part from the other two.
* Swarm control involves a second intervention at the appropriate time in the development cycle.

Next step

The next chapter looks at how and what bees collect for the colony's wellbeing and survival. Flowers and bees are inextricably linked. Flowers provide nectar and pollen, which the bees use as their food. Bees transfer pollen between flowers, effecting pollination and ensuring the next generation. Bees finding a worthwhile food source return to the hive and communicate its location by dancing.

11

Forage

Forage can mean both food and the act of looking for it. Bees are supreme foragers and find their food primarily in flowers. The urban setting can be surprisingly rich, especially in older suburbs and large parks. Gardens may be tiny but, collectively, they offer a large foraging area for bees. Large cities and towns can also be several degrees warmer than the open countryside and this warmth enhances nectar production in flowers.

Plants provide nectar to attract insects and other pollinators. Pollen is the other attractant for many, including bees. This forms the protein part of their diet and enables them to produce brood food for developing larvae. Nectar is the energy-rich carbohydrate part of the bees' diet.

Bee-friendly plants

Urban areas tend to be rich in early pollen-producing plants, including spring bulbs such as aconites, *Anemone blanda*, *Chionodoxa*, crocus, snowdrops and *Scilla,* which are all popular in gardens. You are also likely to find winter-flowering shrubs such as shrubby *Lonicera*, *Viburnum tinus* and *Daphne.* The list can be seemingly endless. If you want to provide versions that are useful to your bees, the rule of thumb, in most cases, is to avoid frilly double flowers and stick with those of the single, wild type. If you were a bee, imagine having to burrow down through all those petals!

The early blooms merge seamlessly into some of the early-flowering trees like Norway maple, male pussy willow (*Salix* or sallow), sycamore, horse chestnut and many others. All these plants mean that in some suburbs, particularly the longer-established ones, there is some sort of honey flow whenever the weather allows, between April and September. Remember to adjust these timings for your geographical latitude.

A major feature of many towns with tree-lined avenues and parks is lime trees (*Tilia* spp.). Most of its commonly planted species flower from the end of June to the second or third week of July. Some of the more exotic species can flower into late August.

Climbers on old buildings can also contribute. The most common one is Boston ivy (*Ampelopsis tricuspidata*). The flowers aren't obvious to us but they are to the bees. The latest flowering climber very useful to bees is common ivy (*Hedera helix*). Often, bees can collect sufficient from this for their winter food. Some beekeepers complain that bees cannot use this honey properly in the winter because it crystallizes in the comb, but mine seem to manage on it quite well.

Key idea

Bees primarily obtain their food, consisting of nectar and pollen, from flowers. Urban areas offer a greater variety and a longer flowering period than rural ones. As living creatures, bees also need water. This is carried to the nest in the honey crop but it is used immediately and not stored in the cells.

Nectar production

Bees collect nectar from flowers for their own purposes. Plants want them to do this in order to achieve pollination, seed formation and the next generation. Nectar is an attractant to pollinators. It is produced in the flower's nectaries at the base of the petals. Often, the shape of the flower itself or marks on the petals tell the pollinator where the nectaries are located. In other blossoms, the lines, or nectar guides, can only be seen in ultraviolet light. Bees can see the ultraviolet part of the spectrum and thus pick up these guidelines, which we, as humans, cannot see.

A few plants produce nectar from places other than their flowers. The commonest is the cherry laurel, which has some patches, called extra-floral nectaries, on the underside of the leaves near the base of the main stem. These seem to produce nectar while the leaf is young but cease to do so as it ages. Dark patches on the leaves of field bean plants also produce nectar.

You are likely to become aware of honeydew when cars parked below trees such as limes or oaks become covered with sticky, sweet drops of fluid which, if left, will grow a black sooty mould. This is easily washed off and causes no lasting harm. It comes from aphids and scale insects. They plug themselves into the plant's sap system and the sap is pushed into their mouths under pressure. They absorb the nitrogenous part and eject the sugary bit. This is honeydew and bees can collect it.

All beekeepers should watch the entrances of their hives. However, get too close and you interfere with the traffic flow. You can observe safely from a distance by using a small pair of binoculars. One sign of a nectar flow is increased activity at the entrance, bees landing heavily with extended abdomens and their legs pushed forward to take the shock of landing.

FROM NECTAR TO HONEY

What happens to the nectar the forager collects? The nectar collected on each foraging trip is stored temporarily in the bee's honey crop while being taken back to the hive. Some of the pollen is removed from each crop load by a valve and, when the worker returns to the hive, she is carrying what equates to a fair-sized drip from a tap. The bee can carry up to 100 mg but the average nectar load is about 40 mg. On average, nectar contains about 40 per cent sugar. Bees are rarely interested in nectar that contains less than 10 per cent as the effort required to collect this outweighs the advantages. In Britain, sugar levels in nectar range from 10 to 60 per cent.

The returning forager passes her nectar load to 'receiver' bees within the hive. Nectar has a much higher water content than honey and the excess has to be driven off. The receiver bee exposes the nectar on her tongue, a drop at a time, before returning it to her crop and adding enzymes. Finally, the drop is placed in an empty cell in the comb. The cell is left open and water continues to evaporate from the nectar. The constant hive ventilation, especially in warm weather, further reduces the nectar's water content. On evenings after a good nectar flow, the bees will be at their entrances, facing towards the hive and fanning outwards, working hard to drive out the moist air.

When the moisture level in the stored nectar has fallen to around 18–19 per cent, it has become honey and the cells are capped over with wax. This capping indicates that the honey is 'ripe' and will keep. This means it can safely be extracted from the combs. The bees have effectively converted the nectar into a state where it will be stable and can be stored. For the bees to use it, in effect, it has to be rehydrated.

Key idea

Nectar provides energy-rich carbohydrates. It is transported back to the hive in the honey crop and then turned into honey. The honey is stored in cells for future use. When it is 'ripe' with a water content of around 19 per cent, it will keep without fermenting.

Pollen

Plants produce pollen in their anthers, often more than they need. It is the male element in the plant's sexual reproduction and contains a high level of protein. Pollen grains are transferred to the receptive female stigma of another flower, either blown on the wind or carried by a pollinator. Bees foraging for nectar inadvertently collect pollen on their hairy bodies. When they visit the next flower, some of the pollen is rubbed off, effecting pollination.

Many plant pollinators, like bees, eat pollen to benefit from the proteins. Some plants do not produce nectar but attract pollinators by pollen production alone.

Pollen varies in colour, depending on flower species. Much, but not all, pollen is yellow, and it may range from dark grey to almost white, with red and blue in between. In fact, the colour of pollen seen in the comb and on the bee can be used to identify on which flowers they are foraging. There are charts of the different pollen colours and it is fascinating to watch your bees and work out where they have been.

Bees have several 'built-in' attributes for collecting pollen and carrying it back to the colony. Anthers are located where pollinating insects will make contact with them. The bee's body hairs are branched and readily pick up the pollen grains. All a worker's six legs carry pollen brushes, rows of stiff hairs used to comb the pollen grains off the body hair. The pollen is passed back to the hind pair of legs, which have modifications to pack and carry loads of pollen. One of the joints on each back leg is slightly dished on the outside face and surrounded by a ring of stiff hairs on the other edge. This is known as the corbicula or pollen basket. There is a single stiff hair in the middle of the face. Pollen is squeezed through the leg joint at the base and up into the corbicula where it is packed into a stable mass.

The pollen is naturally sticky and the bee moistens it with a little nectar. This can affect the colour. For example, horse chestnut pollen is terracotta coloured in the flower but looks maroon in the pollen baskets.

Some bees deliberately collect just pollen. Some collect both nectar and pollen and some nectar alone, discarding the pollen picked up on their body hairs. Bees returning to the hive with pollen loads go to areas where pollen is stored. They select an empty or partially filled cell and dislodge the pollen loads into it. The house bees then tamp them down into layers. A strong colony might collect 50 kg (110 lbs) of pollen in a year. In the autumn in particular, the bees cover stored pollen with honey so that it can be used to rear brood in the late winter and the spring, before fresh pollen is freely available.

Bees are generally good tempered when nectar and pollen are being collected. In many areas, honey flows stop abruptly at some point during June/July. This sudden stop can have a temporary adverse effect on the colony's temper. Bees that were good can become 'testy'. If the nectar flow ceases suddenly, give the bees a few days to settle down before opening the hive. In urban settings, where the bees' reaction to your manipulations is a concern, it is best to keep this reaction in mind.

Key ideas

Pollen sticks to the hairs of the bee's body as she forages. It is collected in the pollen baskets on the hind legs and carried back to the hive.

Pollen is stored in the cells next to the brood nest. Nurse bees eat it and use it to produce brood food for the larvae.

How bees communicate

It is obvious that, with all this variety, plants bloom at different times and in many different places. The food source might be a park full of lime trees or just one small daphne bush. One bee may find the park but on her own cannot fully exploit the treasures of nectar it holds. The solution bees have evolved is the ability to communicate with one another.

Just as body language is a big part of human conversation – we use different voice intonations and facial expressions to communicate the subtle nuances of the information being expressed – so bees use their bodies to convey messages about the location of forage (Fig. 11.1).

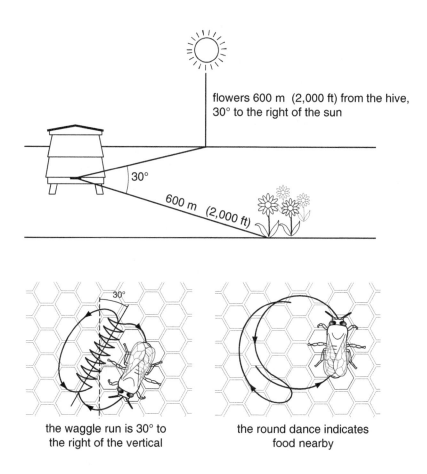

flowers 600 m (2,000 ft) from the hive, 30° to the right of the sun

30°

600 m (2,000 ft)

30°

the waggle run is 30° to
the right of the vertical

the round dance indicates
food nearby

Figure 11.1 How bees communicate

Bees mainly communicate by movements on the combs in the hive that we call dances. The dancing bees tell others the direction in which they need to fly, how far to go and what to look for when they get there.

About one per cent of foraging bees leaving the hive can be described as scouts. Their journey is speculative. They are looking for new nectar sources. Let us say a scout bee finds a rich source of food, both nectar and pollen. She fills her crop and pollen baskets and returns to the hive.

Here, she goes to the main dancing areas on combs near the entrance. Our worker begins by running forward and then describing a semicircle, say, anticlockwise, back to her starting point. Then she vibrates her abdomen rapidly from side to side (the waggle). She then repeats the semicircular run and vibration but this time goes clockwise. The dance attracts other bees, which stop the dancer and are fed some of the nectar from her crop. They now know what they are looking for.

The following bees pick up the scent of the source from the body of the dancer. They emit buzzes that are detectable at a distance through the vibration of the comb surface, thus attracting more followers. Their buzzes make the dancer pause and she emits the same type of vibration.

The waggle and straight run convey information about direction. The bees need a point of reference both inside and outside the hive. Outside, they use the sun. They can determine the point on the horizon directly below the sun's elevated position. In the hive, the direction of the run indicates the direction for the new foragers to fly. If it is vertically upwards, they will fly directly towards the horizon under the sun's position. If it is at 20 degrees to the right of the vertical, it means fly at an angle 20 degrees to the right of the sun. This indication of direction can go round the whole 360-degrees circle. We know that the sun moves across the sky – and so do the bees. The dancers adjust the direction of their dances over time in line with the sun's movement. Now that's clever.

Bees indicate distance by the number and frequency of the waggle phases of the dance. The slower and less frequent they are, the further it is to the nectar source. The richness of the source is indicated by the vigour of the vibrations. The better the source, the more vigorous the waggles.

Potential recruits follow the dancer, picking up all this information.

 Key idea

Bees communicate the type of flowers, their direction and distance from the hive by dancing on the vertical comb in the hive.

The bees have other dances that are seen less often. The main one is the round dance, which does not include a tail-wagging element. This appears to tell recruits to look closely all around the hive in any direction. The dancer first runs in a circle, either way round, and follows this immediately with another circle 'on top' of the first but in the opposite direction. This is the sort of dance bees do when given a feeder full of food. Any syrup spilled around the hive is soon located!

Try it now

* Urban areas can offer a wide variety of forage for bees. Make sure that there is sufficient in your area to support the number of colonies you keep.
* Plant early-flowering spring bulbs and winter-flowering shrubs to provide your bees with nectar and pollen when they most need it.

Focus points

* For a colony to thrive, it must have access to sufficient forage at all times during the active season.
* Flowers attract bees and other pollinators by offering them nectar and pollen.
* Pollinators transport pollen between flowers as they forage, ensuring seed production for the plant.
* Nectar is converted into honey by the addition of enzymes and reduction of the water content.
* 'Ripe' honey is stored in cells sealed with a beeswax capping until it is required.
* Pollen forms the protein part of a bee's diet and is used in the production of brood food for the developing larvae.
* Bees communicate the source and location of forage with their waggle dance.

Next step

The next chapter describes the collection and use of propolis, a resin that occurs naturally on tree buds and bark. Bees collect it and take it back to the hive in their pollen baskets. Here it is used to seal up cracks and cover over foreign objects that the bees cannot or do not want to remove.

12

Propolis

Derived from Greek, the word 'propolis' means 'before [*pro*] the city [*polis*]'. It is a resin that occurs naturally on tree buds and bark. Bees collect a lot of this resin from poplars such as balsam poplars (*Populus balsamifera*) and conifers such as pine trees.

When they are collecting propolis, bees bite off shreds with their mandibles and pass them back to their corbiculae (pollen baskets) on their hind pair of legs. They build up a load that looks similar to a brown pollen load. There is brown-coloured pollen, from white clover for example, but propolis is shiny.

A worker bee can load her pollen baskets with propolis on her own but she needs another bee to help unload them back at the hive.

How bees use propolis

Bees use propolis in two main ways:

▶ **They push it into cracks and crevices inside the hive.**

If you doubt the need to use a hive tool, the sticky nature of propolis and the places where bees deposit it will soon show how necessary it is to have one! Propolis is also used to smooth over irregularities within the hive. One interesting use is at the entrance. Some colonies will use propolis to fill up a large entrance, leaving bee-space holes through which they go in and out. This links nicely with the meaning of the Greek word.

▶ **They use it to cover over anything they find objectionable but are unwilling or unable to remove.**

I have known bees propolize over areas of the broodnest where large patches of larvae had died from chalk brood. Presumably covering the objectionable mass is as good as removing it as far as the bees are concerned. It's a bit like sweeping the dust under the carpet rather than picking it up in a dustpan and throwing it away.

Propolis contains fungicides and bactericides put there by the plant that produced it. These substances perform the same functions in the beehive as they do in plants and humans. This is highly valuable to the bees but I'm not sure that they 'know' this when they collect propolis. They have been known to collect other substances of a similar consistency. On more than one occasion they have been witnessed collecting tar from the road, warmed by the sun. A number of years ago *The British Bee Journal* reported that an Australian tractor factory was in the habit of drying newly painted vehicles outside. At a certain stage in this process, the bees started collecting the paint, to the extent that the tractors had to be painted again!

Harvesting propolis

There is a market for clean propolis. Plastic screens have been produced with v-shaped slits in rows across them. You put them on the hive over the brood box or super and the bees fill the gaps with propolis. To harvest the resin, the screen is put into the freezer for 24 hours, making the propolis brittle. Flexing the screen flakes off the fragments of propolis so they can be collected.

Propolis can be scraped off the wooden parts of the hive but this will always contain contaminants. Collected on a screen, it is clean and can be used for medicinal tinctures. Propolis has been used in traditional medicines for thousands of years. The market for these is stronger on the Continent than in the UK, although you can find propolis remedies in health shops.

A very few people are allergic to propolis. It causes an itchy eczema-like rash that clears up as soon as the beekeeper stops handling combs and frames. Nevertheless, propolis figures today in ointments, shampoos and health treatments. A few people in the UK make the effort to collect propolis but, compared with that for honey and beeswax, demand is small.

Focus points

✳ Propolis is a natural resin collected by bees.

✳ It is used in the hive to fill cracks and to cover over foreign objects.

✳ Bees carry it back to the hive in their pollen baskets. They need another bee to remove it.

✳ Bees will collect materials with a similar consistency, such as melted tar or part-dried paint.

✳ Some people are allergic to propolis.

Next step

The next chapter describes how the beekeeper can remove honey from the hive and how you can make sure that it will not ferment. The stages of the extraction process are outlined, including filtering and bottling. Advice is given on returning the combs to the bees. Honey naturally granulates but this may be with a coarse crystal. A method of producing soft-set honey is described.

13

Harvesting honey

One of the best things about beekeeping is being able to enjoy the honey you have harvested. This is when all the care and attention you have given your bees pays off. The number of times a year you can harvest honey depends on the local conditions and the richness of the forage available to your bees.

Before you can extract honey from the combs, you need to check that it is ready, and then remove the bees from the supers. There are several stages in the extraction process, which must take place in clean surroundings using food-grade equipment. After extraction, the honey must be filtered and bottled.

How bees make honey

As we have seen, bees collect nectar from flowers and carry it back to the hive in their honey crops. During the flight home, the conversion to honey has already been started as enzymes are added to the nectar. Nectar is primarily composed of sucrose, which is a complex sugar or disaccharide. The enzyme breaks some of the bonds in the molecule, converting it into two monosaccharides or simple sugars, glucose and fructose, the principal sugars found in honey.

Nectar also has a high water content. If the bees were to store it just as it comes from the plant, the natural yeasts present would cause it to ferment and turn the solution into alcohol. Now, this is good for humans in that it enables us to make mead, but it is not good for the bees. The next stage in the conversion process is therefore to reduce the water content. At around 18 per cent, the yeasts are prevented from working and the honey is 'ripe' and will keep.

It is said that honey will keep virtually for ever. Honey found in one of the pharaoh's tombs is said to have been edible. Even if this was so, I doubt it would have tasted very nice, as all the volatile flavour compounds would have been long gone.

On returning to the hive, a forager passes the nectar from her crop to a 'receiver' bee, which deals with it as described previously and deposits it in a cell. The bee will eat a small proportion of the nectar as food. This nectar passes through the proventriculus, a four-lobed valve, and into the digestive system.

In the hive, the brood nest has an arc of pollen stored above it so that the nurse bees have easy access to it to produce brood food. The honey is stored above this, as well as around the sides of the brood nest. Full cells are sealed with a wax capping until the contents are required.

Removing the honey harvest

When you want to harvest honey from your hive, you have to remove it – minus the bees.

The first thing to consider is that, by doing this, you are reducing the hive volume. Thus, when you remove a super you need to replace it with an empty one. This is why you need to have three or four supers available for each of your colonies. The exception to this rule is at the end of the season, when the nectar flow is coming to an end and the bees are preparing for winter by packing honey round the brood nest.

USING AN ESCAPE

There are several ways to remove the supers but probably the best, particularly in an urban setting, is to use an escape. This is a one-way valve that allows bees to pass out of the super back into the rest of the hive but prevents them from returning.

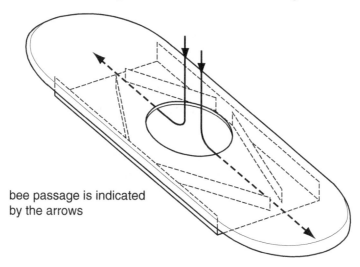

bee passage is indicated
by the arrows

Figure 13.1 A Porter bee escape

The Porter bee escape (Fig. 13.1) is the best-known escape. This comes in two parts. The first consists of an oval plate, designed to fit in the hole in the inner cover. A circular hole in the middle allows bees to enter from the super. The second part of the escape (dotted in the figure) slides under the first. Two pairs of flexible metal springs have one end of each spring attached to the centre of the sides of this lower plate. At the other end, at the outside edge of this plate, the two springs of each pair come together to within a bee space.

The escape is placed in the inner cover, transforming it into a clearer board. The hole goes uppermost, adjacent to the super to be cleared. Worker bees leaving the super are forced to exit through the hole. They then push their way between the ends of the springs on one side or the other and continue into the box below.

If you put a clearer board containing one or two Porter bee escapes on your hive in the morning, you should be able to remove the super(s), clear of bees, in the evening. That said, no mechanism is infallible and your super may not be completely clear. However, these few remaining bees should not cause you problems.

The Canadian is another escape design, which works rapidly. This works on the same principle of allowing bees out of the super but not back, but it employs a narrow tunnel rather than springs. Strips of wood a bee space high are fastened to the underside of the clearer board. They are arranged in a lozenge shape with the widest area under a hole in the board and the ends narrowing towards the board's edges, again to a bee space.

Some close mesh is pinned to the wooden strips, which allows the odour of the bees in the brood box to reach those in the super. Bees leave the super and are channelled into the lower box through the gaps between the wooden strips. They generally do not find their way back, although they may do so if you leave the clearer board on for a long time. A plastic lozenge variation is available commercially and this is just pinned in place below the hole in the clearer board.

Key idea

Before you can extract your honey from the combs, the supers need to be cleared of bees. This is achieved by using a one-way valve, which allows bees to leave the super but not return.

TESTING FOR RIPE HONEY

You may find that a super has not cleared when you return. This could be because it contains a large proportion of uncapped cells containing unripe honey. To test whether uncapped honey is ripe but not yet sealed, take the frame, hold it horizontally over the hive so that drips don't fall outside and attract robber bees, and give it a sharp downwards shake. If a lot of fluid flies out, the honey is not ripe and should be returned to the hive. If only a few drops are ejected, then it is pretty safe to extract. If a bee gets stuck between the springs, that will also prevent the super clearing! If you leave the escapes on the hive for any length of time, bees will start filling the gap near the spring attachment point with propolis, reducing the flexibility of the metal spring. Check that this has not happened before you put the escape in place.

EXTRACTING HONEY FROM THE COMBS

Having cleared the bees from the supers, the next step is to extract the honey from the combs (Fig. 13.2).

Legislation on food hygiene requires you to extract your honey in clean surroundings to which bees, cats, dogs and other pets do not have access. If you only have a few supers to deal with, the ideal place is the kitchen, but get the cook's permission first!

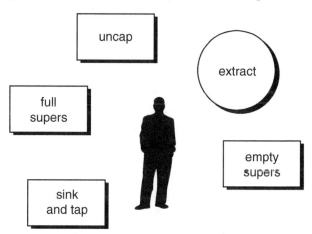

Figure 13.2 A practical layout of the honey extracting area

As well as clean surroundings, you will need two containers of water – the first in which to wash utensils and equipment and the second for washing your hands. It is easiest to use the sink for one and a bucket of water for the other, refreshing this as required.

Arrange the stages of the extraction process so that each is close to the next one. You will have a pile of full supers. The next stage is to uncap the combs over a bowl or other container (Fig. 13.3). The frames then go into the extractor. When they have been emptied, they need to go back into an empty super, ready to be returned to the apiary.

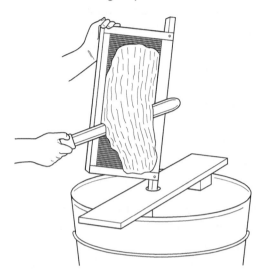

Figure 13.3 Uncapping a frame

Uncapping frames can be messy. In fact, the whole extraction process tends to get very sticky so I recommend you spread newspaper over the floor before you start. This makes it much easier to clear up any honey drips. If honey gets spilled, remove it with an almost dry cloth. A wet one simply spreads the honey over a bigger area, giving you an even larger sticky patch.

1   The first task is to remove the wax capping. For this you will need a knife with a slightly flexible blade longer than the depth of the frame. A serrated edge works well. Find a washing-up bowl or similar and a piece of wood (approximately 50 cm × 25 cm or 2 in. × 1 in., in cross section) that is longer than the bowl's width. Cut a depression in the centre of one face of the wood, big and deep enough to restrain the frame lug. On the opposite face, cut two grooves that will fit over the rim of the bowl.

2   With this wooden bar in place, place the lug of the first super frame in the central depression. Leaning the frame so that the cappings fall away into the bowl, cut across the comb

until the cells are open. It seems to be a natural action to cut upwards. If you do, take care that you don't add some of your finger to the cappings!

3 Combs are not always even. Cut off any bits sticking out, leaving the surface as level as possible. It is easiest to cut to the level of the frame itself. You can retrieve the honey in the cappings later. When the bees use the frames next year, hopefully they will draw the comb more evenly. Cutting the combs level also means they can be replaced in the super in any order. If there are depressions in the comb, just make sure that the cappings have been broken open.

USING AN EXTRACTOR

If you are just starting out, you probably don't want to invest in an extractor. Your local association may have one that is available for members to use. When I first started, a good beekeeping friend let me use his after he had finished extracting his honey crop. This was doubly kind because a proportion of the honey from the first super sticks to the walls of the extractor and you don't get all of it into your bucket. However, that's not the case with subsequent supers.

Extractors must be made from food-grade plastic or stainless steel. They use centrifugal force to spin the honey from the cells. The uncapped frames are loaded into a cage, which rotates on a central spindle inside a drum. The honey is drawn off via a tap at the bottom. The cage is rotated either by hand or with an electric motor.

There are two types of extractor – tangential and radial:

▶ **The tangential extractor** (Fig. 13.4) has a rectangular cage with the frames placed against its mesh at right angles to the radius of the barrel. As the cage rotates, honey is flung from the outside faces of the combs, hitting the barrel and running down to the reservoir. Honey on the inside face of the comb is pushed against the midrib. Therefore, to extract from both sides, you need to turn the combs through 180 degrees so that the face previously facing inwards is now on the outside. This process is repeated until the comb is empty.

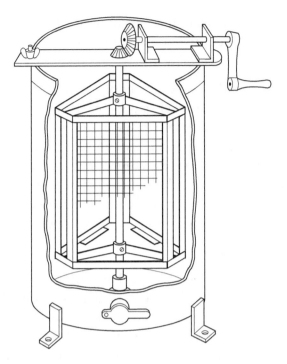

Figure 13.4 A tangential honey extractor

▶ **The radial extractor** (Fig. 13.5) has frames placed along the radii, like the spokes of a wheel. When the cage is rotated, honey is spun out of the cells on both side of the comb at once so there is no need to turn them part way through.

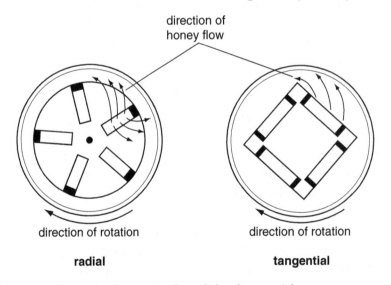

Figure 13.5 The mode of extraction for radial and tangential extractors

Remember this

Do not be greedy and try to get every last drop of honey from the combs because this will risk the comb breaking up. If this happens, you will have a lot of wax to filter out of your honey and the bees will have to rebuild the comb before they can store in it again.

When loading the extractor, try to balance it by placing combs of similar weights opposite to each other. This is not essential but doing so reduces the wobbling that happens until the load naturally evens itself out.

Some people bolt their extractor to the floor to prevent it 'walking' while in operation. This seems the obvious thing to do but it actually puts a great strain on the barrel. An engineer advised me to make a T-shaped frame from reasonably substantial timbers, fix a caster under each of the three ends and then bolt the legs of the extractor stand to the top. He said this would allow the extractor to move and take the strain off the barrel. I was very sceptical but followed his advice. Somewhat to my surprise, when I started the extractor, it simply danced round in a small circle! I recommend his suggestion to you.

The extracted honey will collect in the bottom of the extractor. Keep an eye on the level and run it off through the honey gate tap before it fouls the bottom of the cage. Open the tap carefully because, if you have a good weight of honey behind it, it can shoot out.

HARVESTING WITHOUT AN EXTRACTOR

If you cannot get access to an extractor, you can press the honey from the combs by hand. Cut the comb from the frame, wrap it in muslin and then press it between weights or in a honey press to force the honey out. The cloth will filter the honey. If you decide to take this route, fit your super frames with unwired foundation. You can pull the wires out if the foundation was wired, but unwired makes life a lot easier.

CLEANING UP COMBS

The extracted frames are returned to an empty super at the end of your 'production line'. The combs still contain a thin layer of honey. The best way to clean them up is to get the bees to do it. Take the supers to the apiary in the evening after foraging has finished for the day and place them under the roof but over the inner cover so that the bees can access them. They will lick out the cells and then you can store the supers safely until next year.

To reduce the potential spread of any disease within the apiary, super frames should ideally be returned to the colony from which they came. If you can't do this, at least keep them in the same apiary.

Filtering the honey

After extraction, the next stage is to filter your honey to remove bits of wax and other items like the odd bee's leg that might have crept in. You can filter the honey as it runs from the extractor or you can collect it in a bucket and then filter it into a settling tank.

Note that any buckets you use for handling or storing honey must be made of food-grade plastic. You can buy suitable buckets from equipment suppliers or save ones that previously contained foodstuffs – thoroughly washed, of course.

Stainless-steel filters are available that hang on the honey tap or sit across the top of a bucket. These are coarse filters and will not remove small bits. To remove these, you need to use a fine-mesh filter cloth. The other problem with tap strainers is that they get clogged relatively easily and so need cleaning out regularly.

Settling the honey

A settling tank must again be made from food-grade plastic or stainless steel, with a honey gate right at the bottom edge so that all the honey can be run out. Our settling tank has a separate top section with a coarse filter across the bottom. This is useful to support the filter cloth, which is secured firmly across the top. Use some string or even a bungee cord that pulls really tight. Allow the cloth to sag so that it rests on the metal filter. If you do not have this arrangement, allow your cloth to drape inside the tank. If you stretch it tightly across the top, the weight of honey can pull the cloth away from the string holding it in place.

Very importantly, first check that the tap on the settling tank is closed. Honey flows completely silently and you may only realize your error when there is a pool collecting on the floor!

Take your bucket of honey from the extractor and pour it carefully and smoothly into the settling tank. Running it down the side helps to reduce the impact on the filter cloth. As with the tap filters, keep an eye on the filter cloth and take it off and wash it out if the honey is passing through slowly. Wash it from the clean side to avoid getting pieces of wax on this side and hence into your honey when you put it back in place. You can hang it on a washing line and use a fine spray from the hose to remove the bits. Dry off the excess moisture before putting it back on the tank. It is good to have two cloths that can be alternated.

Bottling the honey

With your honey in the settling tank, you can now bottle it immediately or run it into buckets, let it set and deal with it later.

Standard design squat honey jars are available from equipment suppliers. Carriage is expensive, so plan ahead to collect your order if you visit one of the major beekeeping events. Your association may well make a bulk order and you can buy small quantities from there.

The regulations have changed and honey can now be sold in any weight. However, I suggest you to keep to the usual 1 lb, ½ lb or 12 oz (454 g, 272 g or 340 g) as these jars are readily available and, more importantly, so are the labels that conform to current legislation. Of course, this is only important if you are thinking of selling your honey. If you are only using it yourself or giving it away to family and friends, then you can use whatever containers you like.

1 Whether you are bottling direct into jars or filling buckets, take care when you open the settling tank tap. With a good weight of honey in the tank, especially if it is warm and runny, it can shoot out with some force and at quite an angle. Slowly, slowly is good.

2 Close the tap before the honey reaches the top of the jar as there is always some still to drop inside. It can be topped up to the final level with short, small tap openings.

3 When your jar is filled, put on the lid straight away so that no dust gets in. Such particles act as centres of crystallization and can start the granulation process (see below). If you are filling buckets, leave a gap at the top. Expel excess air when you fit the lid.

Controlling granulation

Honey naturally granulates, or crystallizes. It is a super-saturated solution of glucose in fructose and the glucose precipitates as crystals. The crystals can vary in size from small and fine to large and coarse. The former is obviously pleasanter on the palate than the latter but, if you just leave your honey to set, you cannot predict which way it will go.

Virtually all honeys will granulate over time. The well-known exception is acacia, but you don't get big crops of that in the UK. Honey can also be kept liquid for longer if it is heated to a high temperature but this is not recommended because it not only destroys the volatile compounds which give different honeys their unique tastes and aromas but it also raises the level of hydroxymethylfurfural (HMF). Unless you are going to sell your honey, when your honey has to comply with legal limits for the HMF level, this need not concern you. However, you really want to keep your honey as near to how it came out of the hive as possible, so excessive heating is not recommended.

PRODUCING SOFT-SET HONEY

The set honey you buy in supermarkets can be scooped out of the jar easily and has a smooth texture. How do they do this? They control the granulation and you can do this as well, producing what beekeepers know as soft-set honey. Others will use the term 'creamed honey' but I avoid this as it has implications that the honey has been whipped. That is the last thing you want to do because incorporating air also incorporates airborne yeasts, which will cause fermentation.

To produce soft-set honey, you first need a seed honey. This is one with the desired soft, smooth texture. You may be lucky and some of your honey may set naturally like this. If not, look for a jar of local honey that fits the bill. The process incorporates one unit of seed honey into nine units of liquefied honey.

1　Place the liquid honey into a suitable container leaving space at the top.

2　Warm the seed honey gently until it is mobile and then add it to the bulk liquid.

3　Stir thoroughly, incorporating as little air as possible. This will be easier if your liquid honey is also warm, but cooler than the seed so that this is not melted. Special 'honey creamers' are available but, for small amounts, use a large spoon. In either case, stir without breaking the surface if possible. You want to distribute the small seed crystals evenly through the liquid. This provides thousands of centres of crystallization around which the liquid honey will granulate. The result is soft-set honey.

You can bottle this straight away or return it to the bucket to set.

When you want to bottle your bulk honey, it needs to be warmed sufficiently so that it will flow. You can make a special warming cabinet where you can control the temperature or, for small amounts, short bursts in the microwave followed by a resting period and a stir will work. Be careful not to overheat your honey whichever method you use. When the honey will flow easily, tip it into the settling tank and then leave it for a couple of hours. This allows any air to rise to the top. Bottle the honey until you start to see small air bubbles coming out of the tap.

Either run the remainder back into a bucket with other honey or bottle it for use yourself. If you bottle soft-set honey directly after you produce it, when it sets you may find that it draws away from the shoulders of the jar or you may see lighter lines within the honey. There is nothing wrong with this honey. Shrinkage can occur as the honey granulates. To rectify this, warm the jar gently with short bursts in the microwave and give it a good stir. It should then set again without shrinking. The lines are known as frosting and are streams of small air bubbles trapped as the honey sets. You can also remove these by gently warming the jar and stirring.

If you leave your liquid honey to set naturally, you may find it separates out with large crystals at the bottom and liquid at the top. This upper layer has a higher water content than the lower one and is therefore prone to ferment.

Cut-comb honey

If you do not want to be bothered with extracting your honey, you can produce 'cut comb'. Cut-comb honey is taken direct from the comb. You can buy plastic boxes in which to put the pieces. To eat it, spread both wax and honey on your toast. However, in this case you need to plan ahead and make sure that you use thin, unwired foundation in your supers. You can understand why you don't want a thick wax midrib or strengthening wires.

Once your bees have drawn out unwired foundation, leave about 25 mm (1 in.) along the top when you cut it out and shape the bottom edge of this into a rough V shape. When you give this back to the bees, they will oblige by drawing out natural unwired comb for your next batch of cut-comb honey.

Monofloral honey

Most UK honey will be multifloral, especially if you take a crop from the hive only once or twice during the season. If you want to produce monofloral honey, you will need to know where your bees are foraging and be prepared to extract individual combs or even parts of these. Where a major forage source dominates the landscape, you will be able to get nearer to production of a monofloral honey. Such crops include oilseed rape and ling heather, both of which are unlikely to be found in an urban situation.

Key idea

Most honey is multifloral. It is not practical to produce monofloral honeys in an urban situation. Honeys produced at different times during the season will taste different because of the different range of plants in flower being visited by the bees.

Selling honey

If you have surplus honey that you wish to sell, you will have to comply with the Honey (England) Regulations 2015 (and similar legislation for Scotland, Wales and Northern Ireland) or, in the USA, with regulations produced by your state. Legislation includes details of the type and composition of your honey and how jars must be labelled. Rather than go into the details here, I recommend you start by purchasing labels from an equipment supplier, which will conform to the regulations.

Honey for sale must carry a 'Best before' date and most beekeepers use one of one or two years after bottling. You must also include a lot number, beginning with L, on your jars and keep records of your batches and where they are sold. You can download guidelines from the Food Standards Agency at www.food.gov.uk and the British Beekeepers' Association (www.bbka.org.uk) also produces a helpful leaflet. In the USA the National Honey Board (www.honey.com) and the American Beekeeping Federation (www.afbnet.org) will be able to advise you.

Focus points

* Use one-way valves to clear bees from the supers so that they can be removed from the hive and the honey extracted.
* To check the honey in unsealed cells in the combs is ripe, hold the frame horizontally over the hive and give it a shake. If only a few drops leave the comb it can safely be extracted.
* Extracting unripe honey with too high a water content means that it will ferment.
* Honey extraction must be undertaken in clean surroundings using food-grade equipment.
* Arrange the stages in the extraction process in order and close to each other.
* You can use a tangential or radial extractor.
* Honey must be filtered after extraction to remove foreign objects such as bits of wax.
* Honey granulates naturally as it is a super-saturated solution of glucose in fructose.
* Smooth, soft-set honey can be produced by introducing thousands of minute centres of crystallization.
* Production of cut-comb honey requires the use of thin unwired foundation.
* Selling honey is subject to legislation, so check that yours complies before doing so.

Next step

The next chapter describes the production of beeswax and its uses in the hive. It also explains methods of harvesting and filtering beeswax.

14

Beeswax

Not surprisingly, only honeybees can make beeswax. All insects produce wax and it is an integral part of their cuticle (part of the exoskeleton), providing a waterproof layer. Bees produce more wax than other insects and its composition is somewhat variable.

Beeswax is a valuable by-product of beekeeping, and it is worth collecting even if you harvest only small quantities. When you extract your honey, you will collect beeswax as you uncap the cells. Once you have washed the honey off the cappings, you can dry and store the wax. You can also harvest beeswax by melting down comb that is no longer needed. Once you have collected enough, you can use the beeswax to make a range of products such as candles and polish.

The role of beeswax in the hive

There is no need for a chemistry lesson to appreciate beeswax and its properties. It is a complex hydrocarbon molecule. Four pairs of glands on the underside of the worker's abdomen produce colourless flakes of wax. These are removed by the hind legs and passed forward to the mandibles where the bee manipulates them and puts them in the right place, forming them into the required shape. Beeswax becomes malleable at 32–35 °C (90–95 °F) and melts at 62–64 °C (143–147 °F).

When bees are producing wax, they hang together in chains. They need to establish a temperature of 35 °C (95 °F) and must also have a good supply of incoming nectar to provide the necessary chemical components.

Most of the beeswax produced is used to make the combs within the hive, while some is used to cap over brood cells or the honey storage cells. The cells are hexagonal and very uniform in size. The indented base consists of three identical diamond-shaped plates. The walls of the cell are then built up vertically from the edges of these lozenges. Workers use their bodies to judge the size of the cells and vibrations to determine that the walls are the correct thickness. The cells are built on either side of a thin beeswax midrib and the bases are offset so that they fit together. The hexagonal shape makes the most efficient use of the available space and requires the minimum amount of material. It is also very strong.

The bees use the cells to rear their young and store their honey and pollen. Slightly bigger cells are built when the colony wants to rear the larger drones. When the colony wants to produce new queens, the bees build special cells that hang down over the face or at the edge of the comb.

Bees also build brace comb between adjacent combs to strengthen the overall structure. Any space within the hive that is more than a bee space wide will be filled with 'wild' comb. You will find out all about this if you ever forget to replace a frame at the end of your colony inspection or you hive a swarm in a box without the full complement of frames.

Key ideas

Bees use beeswax to build the basic nest structure. Combs consisting of hexagonal cells are used to raise brood and store honey and pollen. To produce beeswax, bees hang in chains and raise the temperature to 35 °C (95 °F). Production of beeswax requires a good supply of incoming nectar.

Swarms that have been prevented from flying off to their chosen cavity by bad weather, or those that have taken up residence in a natural cavity, will start to build combs. In the first instance, this can be within a bush or hanging from a tree branch. In the second, the comb will be started at the top of the cavity. The combs are extended downwards as the colony expands and needs both somewhere to store honey and pollen and additional space to rear more brood. The combs are built parallel, although not necessarily straight, and will be a bee space apart in the honey storage area and two bee spaces apart in the brood nest.

As a beginner beekeeper, you probably will not have frames of drawn comb available, at least not in the supers. As you gain experience, you will realize how valuable drawn comb is. As long as it is free from disease, you can give it to a colony and the bees can utilize it for brood rearing or honey storage immediately. If you have to give them foundation, they must use incoming nectar to produce wax and build the comb before they can utilize it.

If you have a limited number of drawn combs and the rest is foundation, put a mixture in your supers. One drawn comb placed strategically above the brood nest will encourage the bees up into the super to start drawing out the foundation in the other frames.

Remember this

Drawn comb is a valuable asset that can be used to advantage in various colony manipulations.

Collecting beeswax

When you extract your honey, you will collect beeswax as you uncap the cells. The cappings will contain honey and wax, so you first need to let the honey drain off into a container. Put the cappings on a grid over a food-grade container and leave them in a warm place. The honey that drains out can be added to the bulk that you run out from the extractor. Certainly at the beginning of your beekeeping, this will be your main source of beeswax. Later, you will also be able to harvest it by melting down comb that is no longer required in the hive.

Initially, you are unlikely to harvest a lot of wax but it is still worth collecting. Use soft water to wash the remaining honey off the cappings, dry the wax and store it in an airtight plastic bag. The time will soon come when you have enough to use for a project so it is worthwhile cleaning and filtering it.

Cleaning and melting beeswax

Before using your beeswax, you need to clean it. Cappings wax will probably be the cleanest and purest that you will collect but it will still need to have bits of honey and dirt removed.

Use clean, soft water to wash your wax. This can be filtered rainwater or distilled water. Beeswax reacts to some of the chemicals in tap water, so using this will spoil it. For the same reason, don't use cast-iron vessels when melting beeswax. With only a small quantity of cappings, you can melt them in the microwave. Add the washed wax to some clean, soft water in a glass bowl. With the microwave on high, give the wax short bursts of heat. There is no need to boil the water as the wax will melt before that happens, at around 63 °C (145 °F). If the water boils, this could cause foaming with a resultant mess if it splashes or overflows. An alternative is to put some soft, clean water into a stainless-steel or aluminium pan and add the wax. Heat gently until the wax melts. Again, don't let the water boil.

When the wax has melted, remove the bowl from the heat and let it cool slowly. The wax will form a solid disc on top of the water and can be removed when cold. Wax contains dirt that floats and dirt that sinks. Both can be scraped off the surfaces of your block. The more slowly you let the wax cool, the more distinct will be the division between the wax and the dirt. You can insulate the bowl with some old towels or stand it somewhere cool.

Filtering the beeswax

The next step is to filter the liquid wax to get rid of all traces of dirt. You will need:

- ▶ filter material
- ▶ an old sieve or tin can with both ends removed
- ▶ a collecting container such as a glass bowl or plastic tub
- ▶ a glass jug for pouring the liquid wax.

The filter material will get clogged with dirt and wax but you will be able to use it several times as the retained wax will melt the next time you pour hot wax on to it. Eventually, it will get too clogged and will need renewing. However, the old one is not wasted: roll it up into suitable pieces and use these as firelighters.

The best type of filter material is 'furry' on one side. For clean, spotless wax that will win prizes at shows, I recommend surgical lint, used furry side up. However, this is expensive. For ordinary filtering, materials such as an old shirt, an old sweatshirt (furry side up), an old sheet or two layers of nappy liner work just fine.

Make sure that your filter is firmly supported before you add the weight of liquid wax. One way to do this is to line an old sieve with the material. This will support it and the liquid wax will pour out in a stream from its lowest point. Another method is to take an empty (clean) tin can and remove both ends. Stretch the filter material over one end and secure it firmly with a rubber band or similar. Place the tin, with the filter at the bottom, over your collecting container. Support it on a couple of thin pieces of wood if necessary but make sure it is stable.

The container can be a glass bowl or I find clean plastic margarine containers work well and give you blocks that are easy to handle and store. If you use glass, wipe a thin film of undiluted washing-up liquid over the inner surface to act as a releasing agent so you can remove your block of clean wax when it has set. In plastic containers, the wax tends to shrink away from the sides and the block can be removed. If it resists, you can cut the container away.

A glass jug or other container is needed for pouring the liquid wax easily. Depending on the size of the dirty block, you may first need to break it up. My favourite method is to put the wax in a strong plastic bag, wrap it in an old towel and then hit it with a hammer into smallish pieces. These will melt more easily than larger pieces. You then place the wax in the jug and melt it as before, over water. Wax and water will pass through the filter, leaving the dirt behind. The wax will float and set into a block over the water and it can be removed and dried. Alternatively, you can use a double boiler with the wax in a glass jug sitting on a cloth or piece of wood in a larger saucepan with water coming to halfway up the side.

Pour the wax carefully into your sieve or tin, making sure that it doesn't overflow. Add more wax as the level drops but watch the rate at which it is filtering. If it is slowing down significantly, let the wax in the tin go through and then renew the filter. If the filter blocks while the tin is full of wax, this will either have to be scraped or melted out, which is a fiddly, messy business.

If you have sufficient wax for your project, such as moulding some candles or making some polish, then you can proceed directly. However, if you only have a small quantity or want to use the wax later, let it set. The block can be stored (dry) in a strong airtight plastic bag or wrapped in clingfilm.

Beeswax can be used for a number of things, which are discussed in Chapter 21. If you have enough wax, you could consider making your own foundation, using presses that have the offset honeycomb cell pattern embossed on the two plates. Making foundation can be rewarding but, if you do not want to go to the bother, most major equipment suppliers will exchange your clean blocks of wax for new foundation. Your association may collect wax from members in order to get a better deal.

Remember this

If you do make foundation, *never, ever* make it with wax recovered from colonies which have either American foul brood (AFB) or European foul brood (EFB). AFB spores can survive the temperature of liquid wax and, if you turn that into foundation, you will simply spread the disease throughout your apiary. The causative agent for EFB is also very resistant. If you are unfortunate and your colonies succumb to either of these diseases, the comb *must* be destroyed.

Try it now

✳ Collect your beeswax even if you harvest only small quantities.
✳ Use it to make a range of products such as candles and polish.

Focus points

* ✳ Bees produce beeswax from glands on the underside of their abdomen.
* ✳ The mandibles manipulate it and form it into hexagonal cells on either side of a beeswax midrib.
* ✳ The bees also use it to cap brood and honey storage cells.
* ✳ You will be able to harvest beeswax when uncapping honey storage cells during extraction.
* ✳ If you wash beeswax in clean soft water and dry it, it can be stored until there is sufficient for your project.
* ✳ Be very careful when heating wax, as it is flammable.
* ✳ Use a filter material that is 'furry' on one side to filter wax. When the filter gets really clogged up, make it into firelighters.

Next step

The next chapter will discuss the need to feed colonies when their honey stores have been removed or fall dangerously low. Bees collect and store honey in order to survive the winter and it is the beekeeper's responsibility to ensure that they have enough. Honey can be replaced with sugar syrup or fondant, using a syrup feeder. Hefting can be used to determine when a colony needs feeding.

15

Feeding your bees

I feel there is an unspoken pact between beekeepers and their bees. In return for taking some or all of the bees' honey, the beekeeper has an obligation to see that the bees never starve.

Bees need to have a reserve of food at all times. This reserve will be different at different times of the year. Naturally, the reserves bees build up from foraging are there to maintain the colony in times of dearth. The absolute minimum reserve would be about 4 kg (10 lbs) of stores. If they reach that level, unless the weather improves enough to allow foraging to recommence, you should start feeding your bees immediately. For winter, I would suggest your bees need at least 20 kg (40 lbs) of stores, if they are of the frugal type. Prolific bees may well need twice this.

Preparing to feed

It is vital to ensure that your hive is bee-tight when you are feeding. This activity can promote strong interest from neighbouring colonies and other thieves such as wasps. Make sure that the only way into the hive is through the entrance and reduce this to a size the resident colony can defend easily. This can be 5–8 mm high by 40 mm long ($^3/_8$–$^5/_8$ in. × 1.5 in.). Bees will use all of a larger entrance, 5–8 mm × 150 mm ($^3/_8$–$^5/_8$ in. × 6 in.), say, but all bees make a better job of defending a small entrance. Use your entrance block and reduce the gap by covering part of it with a piece of plastic or stuffing some plastic foam into one end. It doesn't matter if bees have to queue up to get inside. That is preferable to giving thieves easy access.

Making sugar syrup

Sugar syrup, a solution of sugar in water, is prepared in two strengths: strong or thick for winter stores and more dilute or thin for springtime. However, this isn't critical. Thick syrup may be fed at any time but, for the bees to use it, they may have to collect water to thin it down. In the days of imperial measurements, it was easy. Thick syrup was two pounds of sugar dissolved in one pint of water. Thin syrup was one pound of sugar in one pint of water. With metric measurements, however, things become more complicated.

People get very caught up on how many grams per litre will give the same result. I believe the absolute measure is not so important. Bees deal with nectars of various concentrations with no problem, so they can cope with varying strength sugar solutions. The aim with thicker syrup is to have a good amount stored in the combs for the least effort on the part of the bees. Thin syrup requires much more work as more water has to be removed before the solution is ready to store. From the point of view of feeding a colony, it is the weight of dry sugar in the syrup that is critical. I suggest you measure by volume. To make perfectly acceptable thick syrup, proceed as follows:

1 Pour the required weight of dry sugar into a large saucepan or other container.

2 Level off the surface and note how far this comes up the container.

3 Pour in hot water – boiling if you like – and stir while doing so.

4 Add water until it rises 2.5–3.5 mm (1–1½ in.) above the original level of the dry sugar. Keep stirring.

5 Let the solution settle.

6 Return later and stir it again. If you think it needs more water, add a little or reheat the syrup to help dissolve any remaining sugar crystals.

7 Let the syrup cool.

8 Feed it to your bees. When they have eaten/stored all the syrup, you will know that you have fed the weight of sugar you started with.

You don't need to get every sugar crystal to dissolve. Often a few in the bottom seem to be very reluctant to do so. The bees will cope with these.

Remember this

The critical factor when feeding bees is the weight of dry sugar in the feed.

Feeding the syrup

Syrup has to be presented to bees so they have easy access but don't drown in it. The zeal with which bees drink varies with the concentration of the syrup. Below 10 per cent, they aren't interested. The effort involved in reducing the water content is just too much for the benefit gained. Interest increases with the concentration, reaching a peak when the syrup equals the sugar levels found in nectar.

As the concentration approaches that of ripe honey, bees are still keen to eat it but are much less frenzied. Ripe honey in a feeder is eaten quite calmly. The strange thing is that offering the same honey to the bees in the open air produces a frenzied response. You therefore really need to avoid spilling syrup anywhere near your hives, especially when your neighbours live close by.

When you start feeding, bees will go out looking for the faint scent of syrup and are very likely to end up robbing other hives that are being fed. Put on the first feed at dusk so that the initial excitement is mitigated during darkness. Also, start feeding all the colonies in an apiary at the same time. Feeders can be topped up as required but it is essential to make every effort to prevent a spate of robbing.

Remember this

It is your responsibility to ensure that your bees have sufficient food available at all times, especially after removing the honey crop.

Feeding fondant

Feeding bees fondant or candy has become more popular recently. Commercial fondant is made from already inverted sugar. In other words, the sucrose has already been split into glucose and fructose, which is what the bees do with nectar. It is packed in 2.5-kg (5.5-lb.) blocks wrapped in plastic, which are available singly or in packs of five.

All you do is cut a hole in the plastic on one side, peel it back and place the pack, sticky side down, over the feed hole or directly over the cluster if it is not under the feed hole. The bees convert the fondant into liquid stores and place it in the combs. As soon as one pack is emptied, replace it with a new one until you think the colony has had sufficient.

Using fondant is more expensive than making sugar syrup but it cuts down on the messy preparation of the syrup, the possibility of spillage outside the hive and the excitement that feeding can engender. It needs nothing more than an empty super to surround the fondant pack while the bees eat it.

Fondant is eaten relatively slowly, so start this type of feeding in August. Any unused fondant will keep for next year or for use in the spring. Seal it in plastic and store in a cool place.

Types of feeder

To feed sugar syrup to your honeybee colony, you can choose from four main types of feeder (Fig. 15.1).

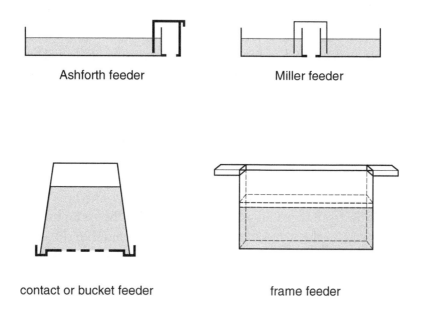

Ashforth feeder

Miller feeder

contact or bucket feeder

frame feeder

Figure 15.1 Common types of feeder

ROUND FEEDERS

These comprise a round reservoir with a lid. A central column/ tube provides access for the bees and a cover over this confines them to a small area around the column. The bees cannot access the bulk of the syrup and should therefore be able to drink safely.

Once emptied, a round feeder can be refilled by simply removing the lid and pouring in more syrup. From the bees' point of view, it is best to feed small(ish) amounts over a period of time. Place the round feeder over the hole in the inner cover (now the feed hole), surround it with an empty super and make the hive bee-tight with the roof.

CONTACT FEEDERS

These can be cheap since they can be made from empty cylindrical food-grade plastic containers. Beekeeping appliance manufacturers supply purpose-made versions, which have a circle of fine mesh in the lid through which the bees access the syrup. For a home-made version, make a number of small holes, 1–2 mm ($^1/_{16}$–$^1/_8$ in.) in diameter, in the lid. If possible, use a container where the lid has a rim approximately the depth of a bee space.

Contact feeders work using atmospheric pressure. Fill the feeder to the top and press on the tight-fitting lid. Have an empty bucket at the side of the hive and invert the feeder over this to avoid spilling syrup on the ground. Hold the feeder level. Syrup will run out into the bucket until a partial vacuum forms over it in the container. The rest of the syrup is then held in by atmospheric pressure.

Place the feeder, holes down, over the feed hole. The bees will suck syrup from these holes. It is essential that bees can move about beneath the lid so, if it does not have a suitable rim, place it on thin pieces of wood. Two old frame bottom bars placed parallel either side of the feed hole work well.

MILLER AND ASHFORTH FEEDERS

Both these feeders have a cross section the same as that of your hive, so they fit on to the hive in the same way as a super. The Miller feeder has a covered central access slot while that for the Ashforth pattern is at one side. These feeders are made of wood and leaking can be a nuisance. Before use, test them by filling the wells with water. If there are leaks, dry the inside thoroughly and paint it with a gloss paint. If there are any large holes, fill them with plastic wood or equivalent before painting.

Both the Miller and the Ashforth feeder can hold most of the colony's requirements for winter feed in one filling, so they may only have to be topped up once or twice.

FRAME FEEDERS

These have the same dimensions as a frame and consist of a reservoir containing a float. This ensures that the bees do not drown in the syrup. Frame feeders are useful for feeding a single colony. Insert them when foraging is finishing for the day and the bees will find the feeder during the night. The other colonies will not realize that feeding is taking place.

Remember this

Check that feeders do not leak *before* you use them.

When to feed

Economical bees, properly fed in the autumn, should not need feeding again in the spring. Prolific bees, descended from those in areas where a more constant nectar flow was available, tend to respond to incoming food by raising brood rather than storing it for winter. You therefore need to monitor the level of stores in your hives when you are feeding them. Prolific bees may need feeding again in the spring.

Feeding can start after the last supers have been removed for extraction. Feeding bees slowly is the best way. They need time to process the syrup/fondant properly so that it will not ferment during the winter. Feeding also wears bees out, so be careful not to feed small colonies too much. You need the autumn bees to go into winter and be available for expansion in the spring. Use a dummy board to restrict a small colony to part of a full brood box and allow them to fill up these combs rather than trying to fill all the combs in the box. In the spring, the dummy board can be removed and the colony allowed to expand on to the empty combs.

Beekeepers with bees at home in the garden can feed little and often until the bees have had enough. Often in the urban situation, small honey flows can carry on right into the autumn, helping colonies considerably. However, these may well be curtailed or be absent altogether, so you must monitor the situation and take action accordingly.

Feeding needs to be complete before the bees start clustering for the winter. In the English Midlands, this is usually early in September. You will have to observe when this happens in your own area. Don't be like the beekeeper who used to contact me every year in October to ask when he should start feeding his bees!

In some years, wasps are a real plague and if they gain access to the hive, they may well steal a great of what is fed to the bees. This may result in an inadequate supply of winter food, again necessitating spring feeding.

Key idea

The short instruction is this: if you think your bees need feeding – feed them!

Hefting the hive

To heft means to lift or hold something to test its weight. Hefting gives you a rough assessment of the weight of your hive, enabling you to check how the colony's winter stores are holding out.

Go to the hive and lift the back of it with your fingers – just enough to slip an imaginary card under the floor. Repeat at the side. The first time you do this, examine the frames and assess what food they contain. One full comb is equivalent to approximately 2.5 kg (5 lbs) of food. Go through the hive and then add the amounts together. Say this is 7.5 kg (15 lbs). Deduct this from the total 20 kg (40 lbs) required (or more for prolific bees) and feed the difference. In this case, it is 12.5 kg (25 lbs).

After you have fed them, heft the hive again. It should feel as though someone has nailed it to the stand. If it doesn't, feed some more. With experience, you will learn the rough level of food in the hive. If you want to be precise, there are devices that will give you an absolute weight.

It can certainly be very interesting to place your hive on a hive scale so that you can monitor weight fluctuations throughout the year. You will be able to see how much nectar has been brought in during a flow and then how much water has been driven off. You will also learn the weight of any swarms you lose!

Do not feed bees over supers unless you really have to. Bees do not distinguish between sugar syrup being fed and nectar. I was once given some honey to taste that was light-coloured, clear and without any honey taste. Bees had moved the results of an over-enthusiastic spring feed from the brood nest to the super to make room for the queen to lay. This is a mistake if done innocently but illegal if you sell the result as honey.

Key idea

You can estimate how much food is in the colony when you heft it.

Try it now

* Be prepared to feed your bees whenever they need it.
* Decide whether to feed liquid syrup or fondant (candy) at the appropriate time of year.
* Feed sufficient in the autumn to last the colony until the spring.
* If you think your bees need feeding, feed them.

Focus points

* Bees may need feeding at any time during the year.
* Liquid syrup should only be fed when the bees are able to move to access it.
* Contact feeders must be surrounded by a super or eke to keep the hive bee-tight.
* Fondant (candy) must be placed over the feed hole or directly over the cluster so that the bees can access it easily.
* Sufficient food should be fed in autumn to maintain the colony until foraging can start in the spring.
* Heft hives to gauge the amount of food stores present.

Next step

The next chapter will discuss ways of preparing your bees for winter, one of the most challenging times of the year. Bees cluster to maintain a core temperature within the hive and use their stores to provide energy for this. When feeding, it is essential that the hive is bee-tight so that no wasps or bees from other colonies can gain access and rob the stores. Hives need to be protected from wind and damp. Colonies must be disturbed as little as possible once the bees have started clustering for the winter.

16

Preparing for winter

Bees naturally prepare for the winter. The stores they have accumulated during the summer are there for them to live on during the period when there is no forage available and temperatures are too low for them to fly out of the hive anyway. When we take their food (honey), we must replace it (with sugar) so that the bees do not suffer from its loss.

Winter is the most challenging time for bees, but there is much you can do to protect them – from the worst of the weather as well as from wasps and bees from other colonies that might try to rob their stores – and help them survive this season.

The bees' honey store

The way bees store honey changes as winter approaches. During the early summer, bees want to store honey away from the brood nest and they use the supers provided. From late summer onwards, increasingly they tend to concentrate their stores around the brood nest. In many areas of the UK, August is a warm month, even hot, but the honey flows are likely to be non-existent. In some urban areas, this dearth will be reduced by exotic nectar sources such as Himalayan balsam and flowers in gardens and parks.

The reduction in foraging either causes or coincides with a reduction in the size of the brood nest. Honey from any late flows, such as the ivy, is stored close to the edge of the reducing brood area. In prolific bees, this is much less obvious and they can have a brood box containing no food at all when the supers are removed. If this happens in your colonies, start feeding immediately. In these circumstances, fondant is a good choice. Feeding syrup would also work. If bees are working well when the weather is good, and if, on hefting, the hive weight is increasing or at least being maintained, there may not be such a pressing need to feed. However, you need to be aware of the situation and act accordingly.

One thing the colony needs in winter is young bees. Very prolific bees produce too many in the autumn at the expense of stores. Much less prolific bees, much more desirable in my opinion, may need the encouragement of an early feed in August. This encourages the queen to keep on laying and the main bulk of the winter food can be given in September. Feeding is best completed by late September in the north and mid-October in the south.

Remember this

Bees prepare for winter by packing honey round the brood nest. Young bees are essential for colony survival as they are the ones who will help colony expansion in the spring.

Key idea

Help your bees prepare for the winter by feeding them sooner rather than later. Feed little and often, if possible.

Clustering

Bees deal with cool temperatures by forming a cluster. In hot weather, they move freely about around their nest. They may even have to fan at the entrance to drive out hot air and allow cool air to flow in. The insulation provided by the hive will help to maintain this comfortable temperature easily during the day but, as ambient temperatures cool, at around 14 °C (57 °F) the bees start forming a loose cluster around the brood nest.

As the temperature outside falls even further, the bees move closer together and cluster shrinks towards its warmest part, the brood area. If there is no brood in the winter, the centre of a cluster will be around 21 °C (70 °F), but in the presence of brood it will be at a slightly higher temperature, 35 °C (95 °F), suitable for healthy development of the brood. If the ambient temperature is −10 °C (14 °F), then the bees can only create heat to maintain the temperature difference by digesting honey and forming an insulating shell of bee bodies around the core. Bees on the outside of the cluster are next to the food and have full crops. However, they may be only just warm enough to stay alive at around −13 °C (8 °F), so there is a constant circulation of bees in the cluster, with the cold, full, ones on the outside that move inside to warm up being replaced by warm, empty, ones from the centre.

The cluster is roughly spherical with the combs enclosed inside it. It has an intimate contact with the food in the frames within it. Some bees crawl into the empty cells within the cluster and, if the colony dies from starvation, these remain behind. Such bees, head first down in the cells, are a telltale sign that the colony starved to death.

Throughout the winter, the colony has to remain in contact with its food. In all the years I have kept bees, I have found that the only way the cluster can move in practice is upwards. In cold spells, even a gap of 25 mm (1 in.) between the outer bees and any new areas of food can be too much. The bees cannot cross it and can die with food just beyond reach. This is 'isolation starvation'. Try to feed enough so that there is a minimum of 50 mm (2 in.) between the outer edge of the brood nest and the top bars.

This is easy, almost automatic, when feeding a full colony. With small colonies, fewer combs and some dummy frames will restrict the bees to a limited number of frames, which they can fill properly rather than trying to spread out food in a complete complement.

Key idea

Bees 'circulate' in the winter cluster, colder ones on the outside moving inside and being replaced by warmer ones from the centre.

During winter, bees eat comparatively little of what has been fed, maybe only 250 g (9 oz) per week. The purpose of most of the autumn food is to help the colony expand in the spring. The space created by eating the food consists of cells where the queen can lay.

Remember this

Do not disturb the winter cluster unless it is absolutely essential.

Defences against robbing

Syrup is an energy-rich food beloved by bees. It is also beloved by wasps. In the autumn, wasps attempt to rob hives that are being fed as well as those that are not. It is a kind of duel – bees versus wasps – and it has probably been going on for millions of years.

Adult wasps receive some or all of their sweet food from their larvae after feeding them with a soup of insect protein. This supply dries up later in the summer when the nest size reduces. The hungry adults turn to other sources and honey stored in a beehive is one of them. We cannot prevent the robbing of hives altogether but we can help our bees hold their own by following these rules:

1   Make sure that wooden feeders do not leak.

2   Don't spill syrup in the apiary.

3   Reduce hive entrances to the point where bees need to queue to get in.

All these defences against wasps are defences against other bees as well. Wasps can steal all the food from small colonies. Other colonies of bees are just as bad – and can probably do it more quickly. It is easier to take precautions to never allow robbing to start than to stop it once it begins. Make sure that the entrance is the only way into the hive and that the colony can defend it.

Feeding might well stop in September but the robbing may continue until much later in the year. I have seen wasps enter a hive as late as 20 December and have heard stories of wasps surviving much later in recent warm winters.

Bees will naturally defend the hive entrance. Once a wasp is inside, however, it can find honey and eat its fill. I once inadvertently trapped a couple of wasps inside an observation hive being taken to a country show. It was interesting to see that the resident bees paid no attention to the intruders. The worst treatment they got could be best described as 'jostling'. Maybe the wasps picked up the hive scent and were therefore accepted as part of the colony?

Securing hive stands

Make sure that your hive stand is placed on a firm surface. This will be fine if you have your hives on a roof but, if they are in your garden or allotment, placing a paving slab down first will provide a good base and prevent the legs sinking into the ground if it becomes soft with rain. Make sure that the slab is level before positioning your hive.

I have seen hives standing on house bricks in an area of bare soil. After a severe thunderstorm, soil was splashed about 25 cm (9 in.) up the sides of the hive. A sturdy stand with legs about 30 cm (12 in.) high positions the hive above such potential sources of damp.

Protecting the hive from wind and damp

As already discussed, it is essential that hives receive some degree of protection from the wind, especially in the winter. Avoid placing hives where the prevailing winds will blow into the entrances. You can reposition a hive by small movements every few days if necessary.

Reduce the amount of damp getting into the hive by using suitable hive stands and protection from wind and rain. In particular, make sure that the roof is rainproof. Once you have done all you can to reduce external damp problems to a minimum, you must then consider internal damp within the hive.

Bees breathe oxygen, as we do, and exhale damper air containing higher levels of carbon dioxide. This air is warmer than the hive's ambient temperature. It will condense on the inner walls of the hive. Bees can be short of water in the winter and I have seen them drinking this condensation. However, damp can promote mould growth. A gentle, steady flow of air through the hive seems to help. An open-mesh floor should solve the problem. If your floors are solid, however, you can still achieve this quite simply, as follows:

1 On a warm day around late October, remove the inner cover.

2 Place a piece of wood, such as a matchstick, across each corner.

3 Replace the inner cover and the roof.

Moisture rises and will escape through the gap you have now created. Carbon dioxide is heavier than air and will exit via the open-mesh floor. Do this late enough in the season and the bees will not propolize the gap. At the end of the winter, around February as far as the bees are concerned, remove the matchsticks and cover the feed hole to help contain the warmth from the bees within the hive.

Winter activities

Winter is the time when beekeepers can leave their bees well alone and take a break! Every few weeks, make sure that your hives are still upright, that entrances aren't blocked by snow or leaves and that the roofs have not been blown off. In short, keep your eye on them in case anything goes wrong.

The bees' winter activities are governed by temperature. They have a 'threshold' temperature of around 8 °C (46 °F). If the hive, particularly the entrance, is in the sun, at this temperature or above, bees may well come out and fly. They may take a 'cleansing flight' to excrete in the air, away from the hive. The longer the bees are confined in the winter, the greater their need for cleansing flights. If the warmth is great enough for long enough, the bees may forage. This is where urban areas have the advantage over rural ones because of the greater likelihood of a wider period of flowering for nectar-bearing plants – in particular, winter-flowering shrubs such as *Viburnum tinus* and *Lonicera* and other plants such as hellebores. Most early bulbs have something for bees in the form of pollen and nectar, so planting these in your borders will be very beneficial.

Winter is the supreme testing time for bees. In spite of the various efforts designed to help them live, bee colonies can fail and die. One of the most common causes of winter deaths is starvation. The next is probably loss of the queen, the third being diseases. These include not only the very obvious ones like the varroa mite but the old retainers such as *Nosema apis* and *Nosema ceranae*. Many bee diseases shorten the lives of the workers. By the spring, when the colony is faced with the extra stress of rearing the new population, that colony fails.

The final straw can be the weather. Poor spring weather, with cold, snow and rain, can finish off many colonies. After some bad winters, losses can average 30–40 per cent. In dry, sheltered apiary sites, losses will be much less. In a 'good' winter, 10 per cent may die, on average, and many beekeepers will lose no colonies at all. Some unlucky ones may lose all their bees. That is the way of averages.

To sum up winter preparation:

▶ Start feeding early and feed plenty.

▶ Make sure that the hive is bee-tight, that is, the only way in is the entrance.

▶ Raise the hives on stands at least 30 cm (12 in.) high.

▶ Ensure that there is gentle ventilation through the hive, raising the inner cover in late autumn if necessary.

▶ Protect hives from wind.

▶ Place hives where they are exposed to the winter sun.

▶ Reduce disturbance to the absolute minimum, carrying out only absolutely essential activities.

Remember this

Winter is the real testing time for bees, especially if it is followed by a poor spring.

Try it now

❋ Complete feeding by late September in the north and mid-October in the south.
❋ Make sure that there are food stores in the combs above the cluster.
❋ Take steps to prevent robbing.
❋ Check that hive stands are secure and stable.
❋ Protect hives from the prevailing wind.
❋ Ensure that there is sufficient ventilation through the hive to prevent damp.

Focus points

* ✻ Bees prepare for winter by packing food round the brood nest.
* ✻ They form a cluster when ambient temperatures are around 14 °C (57 °F)
* ✻ The cluster contracts as ambient temperatures fall so that a constant temperature is maintained at the centre.
* ✻ The stores provide the energy to maintain the cluster temperature.
* ✻ It is easier to prevent robbing than to stop it once it starts.
* ✻ Hives must be on secure stands, protected from the prevailing wind and with sufficient ventilation to prevent damp.
* ✻ Bees must not be disturbed once they form a winter cluster.

Next step

The next chapter takes a look at the major diseases that face our bees. Both brood diseases and adult bee diseases are discussed, along with the causes and the available treatments, if any. Notifiable diseases are highlighted, with advice on what action you must take by law.

17

Bee diseases

Just like other living creatures, bees can get sick. It is your responsibility to recognize disease and know what to do to cure or control it. Some bee diseases are endemic and, at low levels of infection, do not cause problems. However, if an infection takes hold, you need to step in.

As a friend of mine says, 'Dead bees gather no honey.' A slight variation is 'Sick bees gather little honey.' Your emphasis at all times must be on keeping your colonies as healthy as possible. Healthy colonies develop strongly and are in a position to gather the maximum honey crop, given the right forage and weather conditions. They also produce a good population of winter bees, giving them the best chance of surviving the cold months. Both of these are also the goals of the beekeeper.

Recognizing healthy bees

Bee diseases can be divided into two groups: those affecting the larvae and pupae, known as brood diseases, and those affecting the adults – unsurprisingly known as adult bee diseases.

The first step in disease prevention and control is to be able to recognize healthy bees.

In a healthy hive an egg should be positioned in the centre of the cell, curved over slightly. As your eye travels towards the centre of the brood comb, you should see eggs in cells next to those containing larvae. After three days, the egg hatches into a larva. This is fed by the house bees and grows rapidly. Just before the cell is sealed, the larva fills the cell. A healthy larva is pearly white and curled up in a C shape (see Fig. 2.1). Like eggs, they should be in curved patches as part of the normal brood pattern. There should be approximately twice as many larvae as eggs.

The sealed brood should dominate the brood nest by twice as many cells as those containing larvae. The cappings of worker brood should be slightly domed and smooth. Drones are larger than workers and their larger cells have domed cappings to accommodate the bigger pupae.

Learn to recognize healthy brood. You should notice this each time you make an examination. Dedicate at least one inspection in the spring and one in the autumn to looking at brood alone in order to check its health status. This is vital. Anything out of the ordinary that raises your suspicions should be investigated immediately. The sooner you recognize disease and take action, the easier it is to deal with.

Key idea

Maintaining strong, healthy colonies is a major factor in helping your bees collect sufficient honey for their winter feed. Dead bees gather no honey and sick bees gather little.

Brood diseases and other problems

Symptoms of the main brood diseases and other problems are summarized in the following table. Strictly speaking, laying workers, a drone-laying queen and chilled brood are not diseases but they are still problems that need attention. The magazine *Bee Craft* has produced *An Apiary Guide to Bee Disease Recognition,* which has photographic illustrations of brood diseases (see 'Further information' at the end of the book).

Disease	Appearance of brood
American foul brood (AFB) (caused by the bacterium *Paenibacillus larvae* subsp. *larvae*)	Cappings are sunken and look greasy. They are dark and may be perforated. Larvae are white to brown and will 'rope' if pulled out of the cell with a matchstick. They dry to a hard scale, which is difficult to remove. Affected pupae can smell.
European foul brood (EFB) (caused by the bacterium *Melisococcus plutonius*)	Larvae are off-white and lie in unnatural positions in their cells. They can appear 'melted'. Some larvae die in sealed cells but most do so in unsealed cells.
Chalk brood (caused by the fungus *Ascosphaera apis*)	The larva dies, usually after its cell is sealed. The larva is hard and chalky. Some develop fungal spores and look grey or black.
Sacbrood (caused by a virus)	The larva dies after its cell has been sealed and it has spun its cocoon. The virus prevents the final moult. There is no known cure. Sacbrood looks similar to AFB but the dried larvae can be removed easily from their cells. The remains can have the appearance of a Chinese slipper with a curled-up toe.
Stone brood (caused by the fungus *Ascosphaera flavus*)	Symptoms are similar to those of chalk brood. The mummies are easily removed from the comb.

(*Continued*)

Disease	Appearance of brood
Laying workers	The workers lay multiple eggs in the same cell. These worker cells have domed cappings as they contain drones. There is no normal sealed brood and much stored pollen. The colony deteriorates as the proportion of drones increases.
Drone-laying queen	When a queen runs out of sperm stored in her spermatheca, she can continue laying eggs but cannot fertilize them. They develop into drones. Worker comb becomes uneven as cells are distorted to accommodate the larger insect.
Chilled brood	Chilled larvae die and turn black. This is easy to avoid unless brood is left exposed for several hours.

The brood diseases you are most likely encounter are the minor ones – chalk brood and sacbrood – and the serious, notifiable ones – American foul brood (AFB) and European foul brood (EFB).

Minor brood diseases

Of the minor brood diseases, you are most likely to see chalk brood. As the fungus *Ascosphera apis* develops, it turns the larva into a white, chalky solid 'mummy'. In the reproductive phase, the fungus produces spores, which turn the mummy grey and then black. The mummies are loose in the cells and easily removable by the house bees. In bad cases, you may find mummies scattered around the floor.

There is no treatment. Mild cases are not terribly detrimental to the colony, but bear in mind that every mummy represents a worker bee that did not develop and is a loss to the colony population and, potentially, the foraging force and the honey crop.

Some colonies appear to be more prone to chalk brood than others and, for severe cases the recommendation is to replace the queen, preferably with one known to be less susceptible. Transferring the colony on to clean comb may also help to reduce the fungal load. A well-ventilated hive also helps to control chalk brood.

FUMIGATING COMB

Clean comb is that with the minimal chance of carrying disease organisms. The category includes comb newly built from foundation and drawn comb that has been fumigated with an 80-per-cent solution of acetic acid. You can purchase 80-per-cent acid for this purpose. It cannot be sent by post so you will have to collect it from the supplier. Some local beekeeping associations purchase in bulk and then break it down into suitable portions ready for use.

Whenever you handle acid, do so in the open air or in a very well-ventilated space. Wear overalls and protective gloves made of nitrile rubber (latex gloves will not give sufficient protection) as the acid is corrosive to the skin. It can cause skin burns, permanent eye damage and irritation to the mucous membranes, so take great care when using it. If you do get acid in your eyes, rinse them immediately and seek medical advice.

To fumigate drawn comb, take these steps:

1 Place the frames into a brood box or super, removing metal spacers and runners, which are attacked by acid.

2 Put the box on a floor with the entrance blocked. Don't place the floor on concrete as the acid will attack this.

3 Place a pad of absorbent material on the top bars. Fume pads are available from beekeeping equipment suppliers. Cotton wool works but it tends to come apart in wisps when you remove it later.

4 For a brood box, pour approximately 120 ml (¼ pint) of acid on to the pad.

5 After adding the acid to the pad, place the next box of frames on top and repeat the process until you have treated everything.

6 Cover the top pad with plastic and pin this down to the sides of the box. Replace the roof. Seal the joints between the boxes with duct tape and leave the pile to fumigate for a week.

7 Then dismantle the pile and put the boxes at angles on top of each other so that the frames can air before you give them back to bees.

Remember this

You can purchase 80-per-cent acid. It cannot be sent by post so you will have to collect it from the supplier. Some local beekeeping associations purchase in bulk and then break it down into suitable portions ready for use.

Do *not* use glacial acetic acid if at all possible as it is very corrosive and represents a real hazard. If you can only obtain glacial acetic acid and wish to dilute it, make sure that you add ACID to WATER. Doing the dilution the other way round can cause splashing of the concentrated acid.

Serious brood diseases

The other brood diseases are much more serious. In fact, they are notifiable in the UK under the Bee Diseases and Pests Control Orders covering the component countries. This means that by law, if you even suspect their presence, you *must* inform the relevant authority. In England and Wales this is the National Bee Unit. In Scotland it is the Rural Payments Inspections Directorate and in Northern Ireland the Department of Agriculture and Rural Development Northern Ireland. In the Republic of Ireland, it should be reported to Teagsc. In the USA, contact the US Department of Agriculture. (There are more details in 'Further information' at the end of the book.)

American foul brood (AFB) and European foul brood (EFB) are not geographical diseases. Their names relate to where they were first identified. If your colony has EFB, it may be treatable or it could be destroyed. If it has AFB, I am afraid it will be destroyed. Compensation is available through Bee Diseases Insurance Ltd. Bee inspectors are also looking for two other notifiable pests that have not yet been identified in the UK. These are small hive beetle (*Aethina tumida*) and *Tropilaelaps* subspecies mites.

Key idea

Registering on BeeBase (www.nationalbeeunit.com) means that you will be notified of an outbreak of a notifiable disease near you and the necessary steps to take. You can also use BeeBase to access your NBU records and reports.

EUROPEAN FOUL BROOD

It is difficult to confirm EFB visually in the field because the symptoms can be confused with those of other brood diseases (Fig. 17.1). However, the infected larva dies in an unnatural position in the cell. It takes on a melted appearance and eventually dries to a brown scale, which can be removed easily from the cell. In a severe infection, the brood pattern becomes patchy as worker bees remove the dead larvae. Some larvae die after their cells have been sealed and the cappings of these will become sunken, in a similar way to an AFB infection. However, the contents do not 'rope' as they do in AFB (see below). The disease can be reliably identified using a commercially available lateral flow device, similar to a pregnancy testing kit.

Figure 17.1 Larvae affected by European foul brood

If your colony gets EFB, there is a choice of treatments, depending on the colony size, the level of infection and the time of year. It may be treated using the shook-swarm method, with the antibiotic oxytetracycline, or it may be destroyed. These options will be fully explained to you by the seasonal bee inspector (SBI) if the unfortunate situation arises. The shook-swarm treatment described below can be very effective, but a cure is not a foregone conclusion.

The bacterium that causes EFB (*Melisococcus plutonius*) does not produce spores but can live on equipment. After treatment of an infected colony, go over the internal surfaces of the hive boxes, the floor and the inner cover with a blowtorch until the wood turns a dark chocolate brown. Hives made from dense polystyrene cannot be scorched but can be disinfected by immersing them in a solution of 0.5-per-cent sodium hypochlorite for 20 minutes. Use a solution of one part household bleach to five parts water.

▶ THE SHOOK-SWARM METHOD

Here, a mass of bees is shaken out of one hive and into another. The process involves shaking them off their infected brood combs on to replacement frames of foundation. You can carry out the procedure when the bees attempt to swarm naturally. Place a queen excluder under the brood box for the first few days after the procedure (see below).

You will need:

▶ a clean brood box

▶ a clean floor

▶ a set of frames with foundation or clean comb

▶ an extra queen excluder.

The method is as follows:

1 Find and cage the queen and put her in a safe place.

2 Move her colony to one side.

3 Place the queen excluder on a new floor on the original site and then add the new brood box containing a full set of frames.

4 Remove two or three of these frames to make a gap.

5 Shake all the bees from the original brood box into this gap.

6 Make sure that all the bees clinging onto the original box, floor and inner cover are also shaken into the new box.

7 Release the queen into the new brood box.

8 Replace the queen excluder over the brood box and add any supers.

9 Replace the inner cover and roof.

If you do this when the bees are preparing to swarm, there is a danger that they will continue their attempts for a while. The excluder below the brood box will prevent the queen from leaving the colony. As soon as you see the brood nest developing, it can be removed. If you include one or two old brood frames in the new box, they can be removed when the brood in them is sealed. The removed brood is then destroyed. The magazine *Bee Craft* has produced *An Apiary Guide to Integrated Pest Management*, which covers this technique (see 'Further information' at the end of the book).

AMERICAN FOUL BROOD

Paenibacillus larvae subspecies *larvae*, the causative agent of AFB, is extremely resistant. The spores can survive the temperature of molten beeswax and even that of boiling water. They can live for at least 80 years and can easily reinfect bees after that time. Currently, AFB has been identified in one per cent of colonies inspected, which is probably an irreducible minimum.

Figure 17.2 Larvae affected by American foul brood (Courtesy The Animal and Plant Health Agency (APHA), Crown Copyright)

AFB is usually seen in sealed brood. After the larva dies, the capping becomes sunken and perforated. It can also become moist and greasy-looking and can be darker than other cappings. As the disease progresses, the brood becomes patchy and, as the dead larvae decompose, there may be an unpleasant smell. At this stage you can perform the 'ropiness test' with a matchstick. Take a small wooden rod (a matchstick) and stir it into some light-brown larval remains. AFB is confirmed if, when the matchstick is pulled from the cell, a string of material comes with it. This string or 'rope' can be pulled out for a few centimetres (Fig. 17.2). (EFB may rope a little but this always breaks up soon after the matchstick has been removed from the cell.)

Eventually, the larval remains dry into a dark scale, lying on the bottom side of the cell. To see into the cells, hold the comb at an angle of 45 degrees with a light source shining into the cells and look down from the direction of top bar. You will see the rough scales in the cells.

You should welcome the SBI's visit. Since AFB levels in the UK are very low, the visit will almost certainly be a positive experience. If, however, the disease is suspected, the SBI will conduct field tests and, if necessary, send a comb for laboratory confirmation.

If AFB is confirmed, the bees are destroyed. All the frames, bees and honey are burned in a deep hole and the remains buried. The hive boxes are scorched with a blowtorch, which has a temperature sufficient to kill the spores. Polystyrene hives can be disinfected as above.

Key idea

Ask the bee inspector to visit if you have any suspicions of EFB or AFB. They will be pleased to come and look at your bees and give advice.

PREVENTION AND CONTROL METHODS

If either AFB or EFB is suspected, in England and Wales, the SBI will issue a Standstill Notice prohibiting the removal of bees and equipment from the apiary. Colonies within a 5-km (3-mile) radius will be inspected. If either disease is confirmed, control measures will be carried out. If, in six to eight weeks, there are no further signs of the disease, the Standstill Notice will be withdrawn.

NBU bee inspectors carry out a risk-based programme of visits, inspecting colonies for signs of pests and diseases. However, they cannot visit every beekeeper each year so you are still the first line of defence and should check your bees regularly for signs of brood diseases. If you ever have any serious concerns about their health, contact your SBI to arrange a visit. It is always better to be safe than sorry. SBIs are experienced beekeepers and can offer other help and advice as necessary.

The main method of spreading foul brood and other diseases is the beekeeper: transferring infected combs, particularly brood combs, between colonies; not keeping bee suits clean; using equipment without sterilizing it; leaving honey exposed so that other bees can collect it; feeding purchased honey, which can carry viable AFB spores; drifting between colonies; robbing; swarms from unknown sources and, lastly but very importantly, purchasing infected colonies.

In an urban situation, other apiaries may well be near yours. If you have a problem, another beekeeper may offer to help by providing a comb of brood. They may also offer you equipment to hive that unexpected swarm. Be very careful before accepting such kind and well-meant offers. You may end up with more than you bargained for, which is something that neither you nor your friend want. If at all possible, try to locate swarms well away from yours and other apiaries until the queen has started laying and you can inspect both unsealed and sealed brood for disease. If all is well, you can then move the bees into your own apiary.

LAYING WORKERS AND OTHER PROBLEMS

Although laying workers and a drone-laying queen are not diseases, they do represent problems that need to be dealt with. They have similar symptoms. Because the eggs laid in worker cells are unfertilized, they develop into drones, which require more space. The comb becomes distorted and the cappings are domed. In both cases, the solution is to give the colony a new, laying queen.

In the case of laying workers, this is not easy and in most cases is unsuccessful because the workers will not want to accept a new queen. It is probably better not to try to save the colony. Take each comb and shake the bees off in front of another colony or colonies. The bees will beg to be let into the new hive(s). This, and the fact that they then form a very small proportion of the colony's workers, appears to remove their urge to lay eggs and they return to their normal activities in the colony. Remember your neighbours and carry out this operation when they are not around.

You need to find and kill a drone-laying queen before introducing a new queen. I recommend that you try to acquire your new queen locally but also make sure that she is good tempered. As the old queen has been producing drones, the workers in the colony will be getting older and less able to tend the brood. A week or so after you have introduced your new queen, check that she is laying a good brood pattern and, if you have another healthy colony, take a frame of sealed, hatching brood from this and place it next to the new queen's brood nest. The workers that emerge will boost the nurse bee population and help the colony expand.

This scenario just serves to illustrate the advisability of having at least two colonies.

Adult bee diseases

The two main adult bee diseases are nosema and acarine.

NOSEMA

Nosema is the primary adult bee disease throughout the world. It is caused by an organism that affects the bee gut. Worker bees clean up mess in the hive by licking and by biting with their mandibles, accidentally eating nosema spores in the process. These inject the cells lining the gut wall, basically destroying it while severely interfering with the bee's ability to digest food, particularly pollen. This can cause dysentery, although colonies can suffer from dysentery without having nosema. The life of affected bees is shortened, affecting the colony's development in the spring. The affected colony either builds up slowly, stays the same size, or it dwindles and sometimes dies.

The majority of apparently normal colonies all carry some level of nosema. There is no approved treatment and, since the disease continues year after year as bees pick up spores from faeces on the combs, you need to change these as often as is practicable. As soon as the weather is good enough for bees to fly from the hive regularly to excrete outside, the symptoms of the infection gradually disappear.

Nosema is caused by *Nosema apis*. However, in 1996, the 'Asian variant' of nosema, *Nosema ceranae*, was discovered and it is now the major species causing the disease. Under the microscope, it is difficult to tell the two species apart. The effect of *Nosema ceranae* is similar to that of *Nosema apis* but it has a very marked debilitating effect on the populations of affected colonies.

Treatment for both species is the same, with the colony being shaken on to clean comb, as described previously. I prefer to unite 'sick' colonies and eventually change the combs of the combined stock. I consider that this reduces the number of susceptible colonies.

Do not let your brood combs get too old because they can harbour disease organisms. Three years is quite old enough and you should then take steps to renew the comb by replacing it with new sheets of foundation.

ACARINE

Acarine, which should more properly be called an infestation, is caused by a mite (*Acarapis woodi*). As it was first identified there, it is also known as the Isle of Wight disease.

The mite enters the first pair of trachea of young bees. Inside, it breeds, eventually causing a complete blockage. The effect on the bee itself seems to be a shortened life, which, in turn, means that the colony is unable to expand in the spring. If enough bees are infested, the colony may die. There is currently no authorized treatment for acarine in the UK. In the USA, the only legal treatment is menthol. Some imported bees seem to be more susceptible to acarine than most of our indigenous bees. With the bees that many of us keep, it seems simpler to let the susceptible colonies die out and replace them with those showing resistance.

However, the disease seems to have lost its virulence and I have had to treat colonies only on two occasions in the past 40 years.

For diagnosis, a low-power microscope is used to examine the breathing tubes (trachea) in the thorax. Previously, identification was on the basis that these trachea were dark rather than pale ivory. We now know that pale-ivory trachea can actually contain plenty of mites.

All beekeepers should be aware of and know as much as possible about bee diseases. It never hurts to keep up with the most up-to-date information. New treatments may well become available, so I suggest that you keep an eye on the present situation and maintain close contact with other beekeepers and beekeeping associations.

Key idea

Acarine and nosema can only be identified only under the microscope, so find out whether someone in your local beekeeping association offers this service.

Robbing and drifting

Robbing and drifting aren't diseases but a major way in which diseases are passed from one colony to another. When bees 'rob out' a dead or weak colony, they are just as likely to pick up AFB, EFB, nosema or anything else along with the honey stores. Robbing is difficult to stop once it has begun, so make every effort to stop its advent.

Drifting is the tendency for the bees from one colony returning from foraging flights to accidentally enter another one, potentially taking diseases with them. Bees tend to 'drift' along rows, carried by the prevailing wind, with colonies at each end of the row 'acquiring' more bees and honey. This means that they fill more supers and bees also seem to be attracted to taller colonies. While you may want a tidy apiary arrangement, bees are confused by regular patterns. If you want a regular arrangement, you need landmarks to help bees identify their own hives. These can be permanent features such as plants and shrubs. On a rooftop, various objects can act as landmarks.

My choice for an apiary layout would be a rough circle with hive entrances facing outwards. For inspections, you stand inside the circle, so there need to be gaps between the hives. You can set hives in pairs, 30 cm (1 ft) apart on stands, with the stands spaced at least 45–60 cm (1½–2 ft) apart.

Try it now

* To maintain strong, healthy colonies, learn what healthy brood looks like so that you will be alerted to anything abnormal that needs investigating.
* In particular, learn the symptoms of the notifiable diseases, American foul brood (AFB) and European foul brood (EFB).
* Be alert to the possibility of spreading diseases by robbing and drifting.

Focus points

* Bees get sick just like any other living creature.
* It is your responsibility to recognize this and take the necessary remedial action.
* By recognizing healthy brood, you will also recognize sick brood.
* Some diseases have external visible symptoms, whereas others do not.
* Any colony that does not develop in the spring should be investigated for the presence of disease.

Next step

The next chapter takes a closer look at a major pest of honey bee colonies, the varroa mite, and the viruses it can transmit while feeding on the brood. Chemical treatments are outlined and the use of integrated pest management techniques is described.

18

Varroa and viruses

Although many people want to bury their heads in the sand when it comes to disease, I find it both fascinating and essential to understand the diseases that may afflict my bees. The more you understand about diseases, the better you will recognize them and be able to deal with them. All beekeepers need to keep abreast of the latest developments in honeybee treatments.

New beekeepers need to accept that varroa mites are present in virtually every honeybee colony in the UK and the USA. Early detection is key to successful management, so start your beekeeping career on this basis and employ the techniques to control them and the viruses to which bees are prone, outlined in this chapter, from the beginning.

The varroa mite

The parasitic varroa mite (Fig. 18.1) was originally identified as *Varroa jacobsoni*, a pest of the Asian honeybee, *Apis cerana*. People in their wisdom move species around the world and it appears that varroa mites jumped ship into colonies of the western honeybee, *Apis mellifera*, when these were moved to various eastern regions such as Siberia. That said, research has shown that the mite afflicting our bees is a new related species, *Varroa destructor*. This is now a major problem facing beekeepers worldwide (at the time of writing it is not yet present in Australia).

Figure 18.1 A varroa mite (Courtesy The Animal and Plant Health Agency (APHA), Crown Copyright)

Varroa was first identified in the UK in an apiary in Devon in 1992. It soon became clear that the mite was already well established and, although efforts were made to stop it spreading, all colonies in the UK were under threat from this parasite. While varroa mites obviously travelled on bees in swarms, the quickest way that it spread was by the movement of infested colonies around the country.

There is no point wasting time speculating how the mite arrived. The fact is that it is here and we have to be aware of it and how to control it in our colonies. There is no chance that it can be eliminated.

Both the adult mites and young feed on honeybee blood (haemolymph). The mites survive the winter living on adult bees or in brood, if there is any present. However, they can only reproduce in brood. A pregnant female mite enters a cell containing a larva just before it is due to be sealed. She hides in the brood food until the cell is capped and then feeds on the developing pupa and starts to lay eggs of both sexes over a period of time. The resulting nymphs pass through various life stages. The first egg develops into a male which subsequently mates with his sisters and, in worker cells, the eldest pregnant female, and often the next eldest, leaves the cell tucked between the abdominal plates of the newly emerging bee. The remaining females are too immature and die in the cell, as does the male. The average number of viable female mites produced in a worker cell is 1.45.

Varroa mites prefer to breed in drone cells, which are capped for a longer period than worker cells. This means that an average of 4.2 pregnant female offspring mites can be produced. As far as the bee colony is concerned, when drone brood is present, the varroa population can develop very rapidly. Surprisingly, a colony of bees can support a very high mite population. The mite is 1.6 mm ($\frac{1}{20}$ in.) across which is equivalent to us being parasitized by fleas the size of a dinner plate.

Remember this

There are excellent advisory leaflets on this and other bee health topics available in the UK from the National Bee Unit (NBU), which is part of the Animal and Plant Health Agency. In the USA the Environmental Protection Agency (EPA) has links to bee protection organizations in the different states. (See also 'Further information' at the end of the book.) Do get hold of as many leaflets as you can. They can be downloaded from the NBU website BeeBase (www.nationalbeeunit. com), which also contains a wealth of further information about bee health and disease.

Viruses

The real problem with varroa comes from viruses to which bees are already prone and that may be present at very low levels in apparently healthy colonies. These are taken up by the varroa mites as they feed and they get back into bees by the same process. Then they prove to be more virulent than they were before. You are most likely to see the effects of deformed wing virus, which is exactly what its name describes. Adult bees emerge from their cells with shrivelled wings. Because they are unable to fly, they cannot take a full part in the life of the colony.

Monitoring mite levels

Monitoring the mite levels in your colonies allows you to treat at the appropriate time. Mite populations build up more quickly in some years than in others and a control programme that worked one year may not be as effective the next. The recommendation is to monitor four times during the season: early spring, after the spring honey flow, at the time of honey harvest, and late autumn. If you think significant mite invasion is taking place, increase the frequency with which you monitor. Ideally, you should monitor all your colonies each time.

The main methods are monitoring natural mite mortality and uncapping drone brood.

Monitoring natural mite mortality requires an open-mesh floor with a monitoring tray that can be slid in underneath. Wax moths (see Chapter 19) love this sort of situation, so you need to keep the tray clean, particularly during the summer, to discourage them from taking up residence. Decide when you are going to start monitoring and clean the tray. The requisite number of days later, remove the tray, examine the floor debris and count the number of varroa mites. Convert this to a mites-per-day figure.

Uncapping drone brood enables you to count the number of mites on the pupae that you remove from the cells. Choose an area of advanced capped drone brood where the pupae have reached the pink-eyed stage. Slide the prongs of a honey uncapping fork below the cappings, parallel to the comb surface, and lift the pupae out in a single move. Note the number of cells that have been uncapped. Count the varroa mites, which are very obvious against the pale pupal bodies. Repeat this until you have uncapped at least 100 drone cells. Work out the proportion of pupae that carry varroa mites. If you have an infestation of over 5–10 per cent, then it is serious and you need to take immediate steps to prevent colony collapse.

Key idea

There is a useful varroa calculator on BeeBase (http://www.nationalbeeunit.com/public/BeeDiseases/varroaCalculator. cfm). Fill in the boxes and it will give you an estimate of the number of mites in the colony and when treatment is recommended.

Treating varroa

Without treatment, an affected colony will last only a few years. Since the advent of varroa, feral colonies have largely disappeared. Those that do occur don't last (although there are reports of some colonies becoming resistant to the effects of varroa). Generally, to survive, bees need treatment.

It used to be easy. Initially, there were two legally permitted products, Apistan® and Bayvarol®, both consisting of plastic strips impregnated with an appropriate pyrethroid miticide. Hanging the strips between the brood combs for six to eight weeks, generally in August or September, served to reduce the mite population significantly. However, varroa has become resistant to these treatments. This means that mite levels have to be reduced in a variety of other ways.

THE IPM APPROACH

This approach has been given the title 'integrated pest management', or IPM, and it is widely used throughout the agricultural community. Put simply, it means attacking the mites in a number of different ways to keep the population in the colony as low as possible. Keeping mite numbers low is the only way the beekeeper can control the effects of viruses.

IPM does not preclude the use of chemical controls but rather integrates them with other manual biotechnical management practices. In the case of varroa control, these include:

▶ removal of sealed drone brood (containing the reproducing mites)

▶ using open-mesh floors so that mites falling from the bees drop through and cannot return to the brood nest

▶ trapping the queen, confining her so that she can only lay on specific frames which are removed once the brood is sealed

▶ the shook-swarm method (described earlier), where mites on specific frames can be removed and destroyed.

In the spring and early summer, given the chance, bees are likely to build drone comb. To take advantage of this, either place one or more shallow frames in the brood box or fit a shallow sheet of worker foundation in a brood frame. In each case, provided the hive is level, the bees will build natural comb in the space provided. At this time of year, this is very likely to be drone comb. As soon as the brood in this drone comb is all sealed, it can be cut off and burned, thrown away or destroyed. Wild birds like eating drone pupae. Varroa feeding on the drone pupae will, of course, die with it. One or two combs like this in the brood nest will help to control mite numbers.

VARROA CONTROL PRODUCTS

Over the intervening years, additional products have been authorized for varroa control in the UK. These include the varroacides Apiguard®, Thymovar® and ApiLife Var®. Don't confuse the last one with ApiVar®, which is not registered in the UK at the time of writing. These are all based on thymol and other essential oils but are not as effective as pyrethroid strips. They should not be used when there are supers on the hive because they taint the honey.

In early 2013 a formic acid-based treatment, MAQS® (Mite Away Quick Strips) was authorized. These strips can be used at any time during the active season. At the concentration used, formic acid will penetrate cappings and kill mites in sealed brood cells.

Oxalic acid, another organic acid, also kills mites but as it cannot penetrate cappings they are only the phoretic ones that are on adult bees. In September 2015 the UK's Veterinary Medicines Directorate authorized the use of the oxalic acid treatment Api-Bioxal®. This is supplied as a powder, which can be applied as a solution or by sublimation. It can be used in summer and winter but must be applied only when there is no brood in the hive. It is now illegal to use all other oxalic acid products unless they are obtained by a vet from a country in which the products are authorized. Details of vets willing to help beekeepers are obtainable from the Veterinary Medicines Directorate.

It is very important that you use only authorized products to treat varroa in your colonies. As you are applying a veterinary medicine, there are legal requirements associated with it. You must, *by law*, keep a record of where and when the product was purchased, its batch number and expiry date, the name, type and amount of product applied, the date of application and when it was withdrawn from the hive, and its method of disposal. These records must be kept for five years. You can be asked to produce them at any time.

Whichever treatment you decide to use, it is extremely important that you follow the manufacturer's instructions carefully. If a treatment is meant to be in the hive for six weeks, leaving it there for 12 weeks does not make it doubly effective. In fact, it actually contributes to the mites becoming resistant to that product, rendering it useless!

Hive cleansers such as HiveAlive™, BeeVital Hive Clean®, Exomite™ Apis and NutriBee® are also available. These are not registered varroa treatments but can be used to help your colonies remain strong and healthy and therefore more able to overcome the effects of the mites.

Remember this

Beekeepers should assume that all mites are resistant to Apistan® and Bayvarol®, although there is some evidence that the pyrethroids can be effective again after several years when other treatments have been used. It is wise to start your beekeeping career employing IPM techniques from the beginning.

Try it now

* Be aware that varroa mites will almost certainly be present in your honeybee colonies.
* Monitor mite levels in your colonies so that you know the appropriate time to treat them.
* Use the IPM approach to keep mite numbers under control.
* Keep up to date with your local situation by registering on BeeBase if you live in England, Wales or Scotland.

Focus points

* Varroa mites spread viruses within the colony and it is these that eventually kill it.
* Monitoring mite levels gives you the best chance of knowing when treatment is needed to keep mite levels low.
* Only use authorized treatments on your colonies.
* Use treatments in strict accordance with the manufacturer's instructions.
* Keep records of any treatments given.

19

Pests

A pest is defined as 'a destructive insect or other animal that attacks crops, food, livestock, etc.'. As far as bees are concerned, the commonest ones are rodents, wasps, other bees, birds, vandals and wax moths. There are three that have not yet been identified in the UK but you need to be aware of them and keep an eye out. These are the small hive beetle (*Aethina tumida*), *Tropilaelaps* spp. mites and the Asian hornet (*Vespa velutina*). If they do arrive, the sooner their presence is identified, the sooner a contingency plan developed by the National Bee Unit (NBU) can be put in place.

Rodents

Mice and other small rodents can get into hives and equipment. They chew wooden equipment, ruining and devaluing it. They are particularly fond of pollen but also eat other stores. They chew holes in the combs and drag in grass and leaves to make their nests. When the colony is active, the bees' stings will put mice off completely. However, in winter, occupied hives suffer as the bees will not break up the cluster and mice can potentially gain access quite easily.

Mice can squeeze through a gap 12 mm (½ in.) deep by flattening their skulls. This means that a 12-mm circular hole will keep them out, as will a bee space-high entrance of 6–8 mm (¼–⅓ in.).

Mice cannot chew through metal so, even if the roof fits badly, storing your queen excluder on top of the inner cover should stop rodent access that way. A shallow floor, 8 mm (⅓ in.) deep, or a similar height entrance in an entrance block is just as effective. If you want a deeper entrance during the winter, you must protect it with a mouseguard. This is a strip of galvanized metal with 9.5 mm (⅘ in.) holes punched in it. It is pinned over the full entrance. Bees can get through the holes easily, as can air, but mice cannot. Mouseguards should replace entrance blocks by late October or early November.

Rats are kept out of the hive in the same way. Being bigger, rats do more damage. If they get into stored equipment, there is a further problem as they tend to dribble urine as they move about. This can pass on nasty complaints such as Weil's disease. If this happens, replace your combs and give other equipment a good scrubbing with products that kill all known germs.

Wasps and robber bees

The battle between bees and wasps has been going on for millennia. Towards the end of the summer, when queen wasps are reducing their egg laying and the sugary reward from the larvae is drying up, wasps look for alternatives and a beehive containing honey or sugar stores fits the bill nicely. Bees can usually cope, although some, particularly smaller colonies, will succumb. However, with a little help, they can cope more easily. Help your colonies by making sure that the only way into the hive is through the entrance. A bee-tight hive is a wasp-tight hive.

Make sure that all hive roofs fit closely and all ventilation slots are covered with metal gauze. Reduce the size of the entrance to one that the resident colony can easily defend. The bees will not mind queuing to get in if it keeps the wasps out. If challenged by bees, wasps will fly away. Wasps trying to gain access will stimulate the guards, which should be able to keep them out. While wasps nests generally die out with the first frosts, they can continue through October and into November. One year, I saw wasps still trying to rob bees in December.

Small bee colonies are the most vulnerable, so restrict their entrances to one bee space (8 × 8 mm or $^1/_3$ × $^1/_3$ in.). I think moving the brood nest close to the entrance also helps because it brings the guards closer to the point of danger. Except in warm weather, even really large colonies of bees cannot adequately defend an entrance the full width of the hive and the full depth of a deep floor. Bees are most likely to defend the middle of the gap rather than the ends, and wasps can slip in at the sides. Some colonies are, however, too small to help. You, the beekeeper, should assess your colonies and unite the weak to the strong (see earlier). Remember to kill the queen in the colony you dislike most. Remember also that the weakest colony goes above the sheet of newspaper.

Bees rob one another, and my experience is that they will all do this, given the opportunity. As with wasps, provided the colony is strong enough and it can defend its entrance, the bouts of robbing will peter out. Honeybees and wasps can wriggle through holes that look unbelievably small. Quiet, silent robbing like this can deprive a small colony of all its stores. Once inside the hive, wasps and robber bees are left alone and are able to fill up their crops. Guards only make a real effort to repel intruders at the entrance. Help them do their job properly.

Avoid exposed honeycombs, sticky equipment, multiple ways into a hive and small colonies. Don't examine colonies during the day, or stop the inspection as soon as bees from other colonies start showing an interest.

Birds

Swallows, martins and swifts are all capable of eating bees, but the number of those taken is low and bees can cope. European bee eaters (*Merops apiaster*) occasionally nest in the UK but at present this is in insignificant numbers.

The bird that probably causes most problems is the green woodpecker. Once these birds learn what beehives are, in the winter they are capable of smashing fist-sized holes through hive walls. They are after larvae. They have a great tendency to attack occupied hives in preference to stacks of unoccupied equipment.

You can protect hives against woodpeckers by wrapping them with small mesh (25 mm or 1 in.) chicken wire through which bees can still fly into the hive. You need a strip 4.5 times the length of one side of the roof and the depth of your hive from floor to roof, plus a few centimetres to fold over the top. Wrap it round and overlap the ends. Bend the surplus at the top over the roof to keep it in place. Leave the mesh in place until March or April. Removed carefully and rolled up, it should last a long time.

Key idea

Prevention is better than cure, so make sure that your hives are pest proof.

Vandals

Sadly, vandals could be more of a problem in urban areas than rural ones. I have known children break into gardens, turn hives over and throw stones at them. Those returning from a night celebrating in the pub could find a beehive an irresistible target.

Paint hives in subdued colours to make them less noticeable. If possible, keep them out of site of any nearby pathway. Bees on roofs are probably safest in terms of potential vandalism, but if anything untoward should occur, there is a danger if hives are knocked down to the street below. If you keep your bees at an out-apiary, ensure that they are secured. Hives can be strapped up so that it is difficult to remove the roof. If such a hive is knocked over, the boxes will be held together and there should be minimal damage to the bees inside. However, this also makes the boxes easy to move and steal. Allotment sites often have security fences but, even so, tucking the hives out of sight reduces the risk of someone deciding that vandalizing them would be good fun. Consider planting a thick (prickly) hedge around the hives, leaving an entrance for yourself that can be secured.

Bee suits in camouflage pattern material will help to make you less noticeable when you are in the apiary, which will draw less attention to its position.

Key idea

Where possible, secure the apiary against unwanted visitors. Place hives in concealed places out of sight of your neighbours or passers-by, or take steps to make them as inconspicuous as possible.

Wax moths

Wax moths are a particular problem in stored comb, although they can also cause problems in weak colonies. The solution to the latter is to always ensure that your colonies are strong. This is good advice for any sort of disease, as strong colonies are much more able to deal with problems than weak ones.

There are two types of wax moth, the greater (*Galleria mellonella*) and the lesser (*Achroia grisella*). The larvae of both eat beeswax and the cocoons and debris incorporated into it. Greater wax moths prefer brood comb while lesser wax moths are more of a pest in stored supers. When greater wax moth larvae pupate, they chew grooves into wooden hive surfaces, leaving dents behind when the adults emerge. Lesser wax moth larvae spin their cocoons in the tunnels that they have made under the cappings. This leaves a trail of webbing, frass and other debris. Both species are a menace and can cause considerable damage to your precious comb.

To deal with affected combs, cut them from the frames and destroy them. Clean the frames by scraping off the wax and propolis, including from the grooves in the side bars, and then scrubbing the frame with a strong solution of washing soda. Fit sheets of foundation before reusing the frames.

To avoid damage by these moths, store your supers in cold places:

▶ Store supers containing empty comb in piles, outside in a corner of the apiary on stands. Protect the pile from mice by putting a mouse-proof mesh (such as a queen excluder) on both the bottom and top of the pile. Cover the pile with a sound roof to protect it from the rain.

▶ Once your colonies are ready for winter, store supers temporarily on top of the inner cover with the feed hole open. Again, complete the hive with a sound roof.

▶ Super combs can also be stored in a deep freeze. This will kill the wax moth eggs and larvae. Get permission from the cook first!

▶ You can place each super of comb in a strong bin bag, close it up and store it in a cold place. It is important to protect stored supers from mice.

▶ Bag the frames and put them in freezer for a week. Leave them tightly sealed in the bag and store it in a mouse-proof place.

Small hive beetle

The small hive beetle (SHB; *Aethina tumida*) is a relatively minor pest of the African honeybee, which has developed behaviours that control the beetle's effects on the colony. It was identified in Florida in 1998 and in Australia in 2002. By 2000, at least 20,000 colonies in the USA had been destroyed.

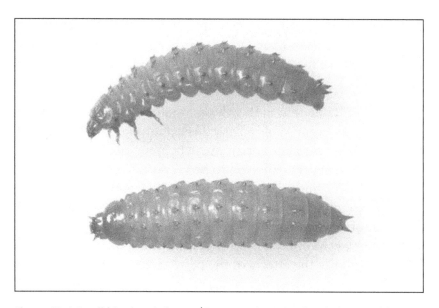

Figure 19.1 Small hive beetle larvae (Courtesy The Animal and Plant Health Agency (APHA), Crown Copyright)

Adults make their way into a hive and lay eggs in nooks and crannies. These hatch in three to six days and the resulting larvae (Fig. 19.1), as well as the adults, defecate in the hive, spoiling both combs and honey, giving all hive parts a slimy appearance. The honey ferments and runs out of the hive. When the larvae mature after 10–14 days, they enter the wandering phase where they leave the hive and search out suitable places to pupate in the soil. Pupation takes 15–60 days and the emerging adults return to the hive to repeat the cycle. Small hive beetles can survive anywhere that bees can live.

In September 2014 small hive beetle was confirmed in the Calabria region of southern Italy and subsequently in Sicily. Inspections within a 20-km (12-mile) radius of the original discovery identified a large number of infected apiaries. The authorities are monitoring the situation.

SHB could arrive in the UK in swarms, packaged bees, used beekeeping equipment, soil around the roots of plants and in fruit. Since 2011, large numbers of package bees (nuclei) and queens have been imported from Italy, although the package bees have come primarily from northern Italy. NBU inspectors have conducted priority inspections of apiaries belonging to beekeepers in England and Wales who have imported bees from Italy. To date, at the end of 2014, all such apiaries have proved negative for SHB.

It seems inevitable that SHB will eventually find its way to the UK. However, because of our northerly latitude, the chances are that it will be a pest but not an overwhelming one. If (when) it arrives, beekeepers will need to be very strict with apiary hygiene, not leaving comb and honey lying around, and will have to extract honey as soon as it is taken from the hive. These are, of course, good beekeeping practices which you should be doing anyway. Maintaining strong, healthy colonies will also help to minimize the effects of the pest. Mechanical methods of control, such as beetle traps hung between the frames or placed on the floor of the hive, are available and new improved designs are being developed all the time.

Because the beetle could potentially cause so much damage to UK beekeeping, anyone discovering it or suspecting its presence is required *by law* to report it to the relevant authorities (see 'Further information').

Key idea

All beekeepers must remain vigilant and keep an eye open for SHB. Early detection by beekeepers could prevent an early infestation becoming an epidemic. A useful information leaflet is available from the National Bee Unit (see 'Further information').

Tropilaelaps mites

Two species of the *Tropilaelaps* mite (Fig. 19.2) have been identified: *Tropilaelaps clareae* and *Tropilaelaps koernigerum*. They were originally a pest of the giant Asian honeybees *Apis dorsata* and *Apis laboriosa*. They infest western honeybee (*Apis mellifera*) colonies in much the same way as *Varroa destructor* but breed about 25 times faster.

Figure 19.2 A *Tropilaelaps* mite (Courtesy The Animal and Plant Health Agency (APHA), Crown Copyright)

Tropilaelaps is probably a very serious pest indeed, and it can be detected in the same way as varroa. It is a notifiable disease and therefore you must, by law, inform the relevant authorities if you suspect its presence. Information and a well-illustrated booklet are available from the National Bee Unit (see 'Further information' for details). There is no doubt that the mites could live in most areas of the UK, surviving the winter in the brood produced by many colonies. Since it eats brood, it can be controlled by caging the queen for a time to interrupt the brood cycle.

Remember this

* Make sure that your hives are bee-tight, which means they are wasp-tight.
* Protect hives against woodpeckers and rodents in winter.
* Maintain strong healthy colonies to deter wax moths.
* Learn to identify small hive beetle and *Tropilaelaps* mites and be vigilant.

The Asian hornet

Possibly an even greater threat to our bees is the Asian hornet, *Vespa velutina* (Fig. 19.3). It is believed to have arrived in southern France from Asia in 2005. It became established very quickly and has spread up the country to the English Channel. It is now also present in Spain and Belgium. It causes many problems for both beekeepers and biodiversity in general. Hornet nests built in the open air can become very large. The worker hornets predate a number of insects to obtain the protein-rich diet required by their larvae. These include honeybee colonies and an attack can cause significant harm.

Figure 19.3 An Asian hornet (Courtesy The Animal and Plant Health Agency (APHA), Crown Copyright)

To have the best chance of intercepting the Asian hornet before it can become established in the UK, it is essential that all beekeepers can correctly identify the species. They must also monitor for its arrival and report any suspicious sightings to the Non-Native Species Secretariat (alert_nonnative@ceh.co.uk). Plans are available on BeeBase for making a hornet trap from an old drinks bottle and different designs are available commercially. A patented trap, ApiShield®, is designed to replace the hive floor and lure hornets inside. They are then trapped and die from dehydration.

Further information and an identification sheet are available on BeeBase. *Vespa velutina* is smaller than the UK's native hornet, *Vespa crabro*. Its thorax is entirely dark, it has a dark abdomen and the tips of its legs are yellow.

Remember this

Learn about the Asian hornet and how to recognize it. Set out hornet traps for an early warning should it arrive in the UK. However, check these regularly so that you can release any European hornets and other non-target insects while they are still alive.

Try it now

* Be aware of the different pests that can cause problems for your bees.
* Take the necessary steps to prevent or deter them.
* Be aware of future threats and keep a look out for small hive beetle, *Tropilaelaps* mites and the Asian hornet.
* Report any suspicious sightings to the authorities.

Focus points

* Pests come in a variety of forms and some can do more damage than others.
* All possible steps must be taken to prevent pest attacks.
* Robbing is easier to prevent than to stop once started.
* Wax moths can destroy stored comb.
* Identifying exotic pests sooner rather than later is vital if attempts are to be made to eliminate or at least control them.

Next step

The next chapter considers ways in which you can learn more about your bees and how to look after them. Your local association is likely to run courses and there is a structured examination system if you want to take your learning further. Many excellent beekeeping books and magazines are available, and just watching others handle bee colonies can also be very educational.

Sources of information and support

When you start keeping bees, you will probably feel overwhelmed by how much there is to learn. After two or three seasons, you might feel you know it all – well, at least the basics of colony inspection and swarm control – and that beekeeping is somewhat repetitive and even a bit boring.

Nothing could be further from the truth. I hope you now understand this and realize that, every time you go to your bees, just how much more there is to know about these fascinating insects. Beekeeping is a hobby where the more you know the more interesting it becomes.

So how do you learn more about bees? There are many ways and you just need to find the ones that suit you.

Local associations

One of the best sources of information is your local association. Membership instantly gives you access to many beekeepers with a wide range of experience. I would be very surprised if you failed to find a beekeeper who did not want to talk about bees or couldn't give you some help. They may not have an immediate answer but they are sure to know somebody who does.

Most local associations arrange for visiting lecturers, particularly through the winter months, and you can gain much by attending these talks. Don't worry if the subject sounds too complicated or advanced; you will always learn something. And there is the benefit of the chat over a cup of tea afterwards.

The majority of associations arrange training courses, including ones for beginners where you spend a few sessions learning the theory and then have a chance to actually look at colonies in the flesh. If you are not sure if beekeeping is for you, I recommend attending such a course before deciding to spend a lot of money on the kit and the bees.

After attending the beginners' course, you may well be linked with an experienced mentor. I can really recommend this. I don't know what I would have done if I hadn't had a good beekeeping friend when I started out.

Remember this

The more you know about how the bee colony operates, the better you can work with the bees to the advantage of both sides. The more you learn, the more you will realize just how much more there is to learn about bees.

Associations also organize courses for those who have kept bees for a few years but want to learn more. These can be practical, where the group visits different apiaries, inspecting the bees and discussing the best courses of action and the reasons for them. Other practical courses can be more specialized, dealing with subjects such as queen rearing or microscopy. Theoretical courses may cover anatomy, bee behaviour and the like.

Key idea

Take advantage of the collective experience available in your local beekeeping association.

LEARNING BY WATCHING

In the summer, associations often hold apiary meetings. These offer an excellent way to learn by seeing different hives, different apiaries and, probably most importantly, different bees. It was only when he went to another apiary that a friend of mine discovered that not all bees follow you back to the house after you have closed up the hive! You can also learn a lot by watching others inspecting a colony. You will rapidly appreciate the difference between those who are good at handling bees and those who are not. You can learn a lot from both.

BEEKEEPING EXAMINATIONS

The British Beekeepers' Association (BBKA) has a widely accepted and admired examination system. This starts with the Basic Assessment, where you inspect a colony and the assessor asks you questions. The ultimate achievement is to become a BBKA Master Beekeeper by studying for and passing a number of specialized modules as well as taking a practical general husbandry examination. Details of the examination structure can be found on the BBKA website. The BBKA also offers a correspondence course if you prefer that.

The ultimate UK beekeeping qualification is the National Diploma in Beekeeping (NDB). This is organized by a separate board (details at www.national-diploma-bees.org.uk).

There is no obligation to take any of the examinations, but I guarantee that studying the syllabi will expand your beekeeping knowledge and understanding. This has got to be good for you and your bees.

Key idea

Studying a syllabus can be extremely beneficial, even if you do not want to take the actual examination.

Books and magazines

If evening classes or structured courses are not your thing, there is a vast library of beekeeping books, a small portion of which is listed in the 'Further information' section of this one.

Beekeeping magazines help keep you up to date with the latest happenings in the beekeeping world. They include a range of articles from those for the beginner to the latest bee research, and everything in between. Many are now available electronically, so you can read them on your phone or tablet wherever you are. Again, details are in the 'Further information' chapter.

Try it now

* Find an appropriate course and learn more about your bees and how to care for them.
* Attend beekeeping meetings and get information from the lecturers.
* Find a bee buddy who can help you if you have problems.
* Utilize the wide range of books and beekeeping magazines available to you.

Focus points

* Learning about bees will make your beekeeping much more interesting.
* Your beekeeping will be more fulfilling the more you understand.
* There are various learning opportunities. Find the one(s) that suit you best.

Next step

The next chapter introduces activities that are associated with beekeeping. These include using honey in baking and making it into mead as well as the various uses for beeswax. Other crafts such as skep making, or collecting honey pots and bee books will serve to widen your interest in bees and beekeeping.

21

Associated crafts and hobbies

Beekeeping covers more than hive inspections and harvesting honey: associated activities range from making honey and beeswax products to collecting bee-related items, from photography to international travel.

Beekeeping has taken me all over the world. I have attended international beekeeping congresses as far apart as Canada and Australia and led beekeeping holidays to places such as Nepal, Chile and Mexico to raise money for a beekeeping charity.

There are so many activities that relate to bees and their products, and they all offer you a way to extend your interest.

Using honey

Let's start with honey. The most obvious application is to spread it on your breakfast toast. However, you can use it elsewhere in the kitchen. As a natural blend of sugars, it can be used as a substitute for cane or beet sugar. However, it typically contains 17–18 per cent water and you must take account of this if you want to substitute it for sugar in a recipe. Just using honey as a glaze on roast meat is fine (don't put it on too early or it will burn) but a straight substitution by weight into a cake recipe will lead to disaster. Assuming that the water content is roughly 20 per cent, or one-fifth, will work well enough.

There are specific recipes for honey cakes. The good ones are designed to produce a moist cake with a definite honey flavour.

Making mead

Honey can also be used to make mead, which is essentially diluted fermented honey. Most commercial meads are very sweet, but you can make dry mead and the best of these can be superb, especially if drunk chilled.

Although honey is acidic, to make a good mead you need to add more acid to the honey–water mix. You also need a yeast nutrient. If you have not made wine before, you might find it easier to start with a melomel, which is a mead made with honey and fruit. Books are available covering the details for making mead and melomel.

Beeswax candles

The other main by-product from your beekeeping activities is beeswax. You are unlikely to collect a lot at the beginning but, if you clean and dry it as described previously, you can store it until you have sufficient for your project.

If you can't wait that long and want to produce some rolled beeswax candles, purchase some unwired foundation. This is available in a wide range of colours specifically for candles.

I recommend buying wick designed for beeswax candles from the suppliers. To burn evenly without flooding (and going out) or guttering (and causing wax to run down the side and make a mess), you need to use the correct wick for the diameter of the finished candle. I have found that a 2.5-cm (1-in.) beeswax candle wick works well for a candle rolled from a sheet of British Standard brood foundation. You can, of course, use any size of foundation but you will have to experiment to determine what size wick you require.

CANDLEMAKING STEP BY STEP

1 Cut a length of wick about 15 mm (½ in.) longer than the short side of the foundation. If the wax is cold and stiff, you can warm it using a hair dryer. Heat it evenly but not too much, or you will end up with a soggy mess of melted wax.

2 Put the foundation on a flat surface with the short side parallel to you. Press the wick into the soft wax about 5 mm (¼ in.) in from the edge. Try to keep it straight. Work along the edge, folding the free end of the wax over the wick a little at a time. Rewarm the wax if necessary and return it to the flat surface.

3 Start rolling the wick away from you so that the foundation rolls up like a Swiss roll. Try to roll it as straight as possible but don't worry too much if you don't manage to keep it dead straight. If you go off course, you can unroll it and start again, but this is usually unsuccessful as the sheet gets stretched in places.

4 Stop rolling about 2–3 cm (1 in.) away from the other edge. Give this an extra warm with the hair dryer (but, again, not too much) and then finish rolling the candle. The warmed edge will stick to the body of the candle and stop it unrolling. Smooth the edge down to make sure that it is all fixed in place.

5 If the base of your candle isn't flat, hold it upright and press it down firmly on to the flat surface. The gaps between the layers allow the 'bulge' to be flattened out.

To make a rolled candle with a shaped top, you need to trim the foundation before you start:

1   Put the foundation on a cutting board (I use an offcut of kitchen work surface).

2   Place a straight edge at one corner of the sheet and the other about 15 mm (½ in.) from the corner at the other end of the long edge. Cut off the thin triangle with a sharp knife.

3   Start rolling from the longest edge and your candle will have a pointed top.

Now you can use your ingenuity and imagination to produce candles of different shapes. You can roll up two different-coloured sheets together. Rolled candles burn for a surprisingly long time but not as long as moulded or dipped candles. You will find instructions for these in books on candlemaking.

Beeswax polish

The simplest recipe for beeswax polish is to shred your block of wax or break it into small pieces and put it in a glass jar with a lid. Add pure gum turpentine to cover the wax. Add the lid and shake the jar every day or so. Over time, the wax will dissolve and you will have a liquid polish. For the economy version, substitute white spirit for the turpentine. This will work just as well because the liquid is merely a carrier to spread the wax evenly across a surface. However, your polish will smell much better with turpentine. If made with white spirit, it will smell like paint!

Heating will dissolve the wax more quickly. Place the turpentine in a glass or stainless-steel bowl that fits over a water bath. Add the wax and put the water bath over a low heat. Stir the mixture until the wax is dissolved. Pour it into suitable containers and let it set.

A solid polish is effective for waxing your furniture but to get the required shine you have to add a lot of elbow grease. You can make a softer cream polish by adding water and soap to form an emulsion. Soap flakes make it easy, or you can grate a bar of soap. You create a 'hot' solution of wax melted in turpentine and a separate solution of soap dissolved in water that is at least at the same temperature. Mixing these and stirring forms the emulsion, which will solidify as it cools.

POLISH RECIPE
You will need:

► 170 g (6 oz) shredded beeswax

► 470 cc (1 pint) pure gum turpentine

► 235 cc (½ pint) soft water/rainwater

► 15 g (½ oz) soap flakes.

1 Prepare more containers than you think you need. Heat the shredded wax and turpentine in a glass or stainless-steel bowl over a water bath.

2 In a separate container, pour boiling water over the soap and stir to dissolve it. This solution can be hotter than the wax–turpentine solution but not colder.

3 Remove everything from the heat. To remove undissolved soap particles, pour the soap solution through a fine strainer and into the wax/turpentine solution. Stir with a balloon whisk.

4 Continue to stir as the mixture cools and, when the polish begins to thicken, pour it into the containers. It is better to have a half-filled container than a congealing mass in the bowl.

Other uses for beeswax

Beeswax has many other uses, some of them somewhat unexpected. These include polishing army boots, waxing pins for lace-making, lost-wax casting and making the mouthpieces for didgeridoos! Beeswax is also used in cold creams, lip balms, hand creams and other cosmetics. If you want to make these, books are available with suitable recipes and instructions. If you wish to sell such cosmetics, you will need to comply with a lot of legislation, including ensuring that all the ingredients you use conform to the appropriate regulations. If you are just making them for your own use, you do not have to go this far but, as cosmetics are applied to your skin, check your ingredients carefully.

There are many other crafts and activities using beeswax, such as encaustic painting, using coloured wax to produce a picture. You can also use beeswax for making models and moulded figures.

Using propolis

The other product of the hive you may wish to utilize is propolis. Propolis has antiseptic properties, which leads to its use in products such as tinctures and throat lozenges. Again, if you want to pursue this, find a book that will give you guidance and recipes. A few people are allergic to propolis and it can cause severe eczema.

Skep making and skep beekeeping

Beekeeping opens doors to all sorts of associated activities. In the old days, bees were kept in straw skeps. Basically, a skep is a domed basket formed from a rope of straw sewn together with split blackberry cane. Skeps were kept in structures known as bee boles (and searching these out is another activity you could pursue). These were designed to protect the skeps from the elements while allowing the bees to fly freely.

Today, skeps are used mainly for collecting swarms. They are light, porous and have a rough interior to which the bees can cling. Some beekeepers maintain colonies in skeps on a permanent basis. This is interesting to do but it has its drawbacks. The first is swarm control, which is particularly important in an urban setting. You will see when queen cells are being built by turning the skep upside down but you cannot follow the usual methods of swarm control, which separate the queen or remove the brood. The only option is to remove the flying bees by moving the skep to a different place in the apiary.

The other difficulty with skep beekeeping is disease inspection. While the combs can be gently eased apart, it is virtually impossible to inspect the brood closely. If you want to try skep beekeeping, ask your seasonal bee inspector for advice first.

Skep making is an art but it is also something that any beekeeper can try. Skilled skep makers offer courses where you can learn the basic techniques. Your skep may not be the desired uniform shape but I bet the bees don't really care about the aesthetics. They are just looking for a suitable cavity, and the various hollows that they choose come in all shapes and sizes. Whatever it looks like, your skep is something you have made and you can be proud of it.

Collecting on a bee theme

Many of us have a collecting instinct and beekeeping gives you opportunities to feed this.

If you haunt charity shops, you will occasionally spot a honey pot. Recently, I came across one in the form of a cork hive similar to those used long ago in mainland Europe. I have also seen plain versions with a bee forming the knob on the lid. Some have the word 'honey' on the side but with others the bee is the only indication of the intended contents.

When 'section' production was common, honey would be placed in special dishes for serving at table. Many of these dishes were elaborately decorated so you can keep an eye out for them as well as fancy honey pots. A search on eBay is also likely to bear fruit.

To follow up this interest, I recommend John Doyle's book, *Collecting Honey Pots* (BBNO, 2009).

Another area in which to indulge your collecting instinct is bee books. Ever since the printing process was invented, we have written books about bees. They contain a wealth of information on the history of beekeeping. You can read about when it was believed that the colony was led by a king surrounded by male soldiers. Then there are all the different methods of managing bees and the numerous hive designs, old and new. You can read about bees being kept in hollow logs and clay pipes, the discovery of the bee space and the resulting development of the movable frame hive, together with new discoveries about bees, their behaviour and their diseases.

There are books about different beekeeping traditions around the world and specialist books on queen rearing, swarm control, beeswax and propolis, to name but a few. The list is virtually endless. My collection numbers over 2,000 titles and this is only a small proportion of what is available. Some are valuable but most are not, so you don't need to break the bank to start your collection.

Bee photography

With the advent of digital and phone cameras, the world of photography has been opened up to almost everyone. This has led to a huge increase in the numbers of photos of bees, beekeepers and associated topics. You can enjoy your bee photography on all sorts of levels, from taking shots of local association meetings to macro close-ups of the bees in the hive and when out foraging.

I would encourage you to record your activities and those of your bees in photographs. By capturing a moment in time, you can study it later and learn more about your bees.

Beekeeping holidays

Another ancillary activity is travel. I was fortunate enough to win a beekeeping holiday to Thailand and this opened my eyes to the fact that there are honeybee varieties other than the ones in my hives. In fact, Asia has a far greater diversity of honeybees than we do in the West. There are big ones and small ones, those that nest in cavities and those that nest in the open air. Beekeeping traditions vary around the globe and watching honey hunters removing combs of the giant honeybee from high up on a cliff in Nepal remains one of my most exciting experiences. In the Tropics, there are even bees that don't sting – although they can certainly defend themselves in other ways.

If you want to combine travel with your interest in bees, look out for beekeeping holidays. These are often organized by beekeeping charities, so you know your trip is also contributing to helping fellow beekeepers in developing countries.

Honey shows

Your local association may well hold a honey show and this can be another 'added interest'. Shows allow you to compare your honey and hive products with those of others and it is a real thrill to see that First Prize card against your exhibit.

Honey shows are not just about honey. A number of the activities described in this chapter are included in the competition classes at honey shows, so you can hone your skills and enter these. Shows often include classes for things like honey cakes, honey preserves, beeswax blocks, candles, polish, sweet and dry meads and metheglin. Other classes may include photography, educational displays, needlework and other art.

There is not space here to go into detail about showing. Suffice to say that the winning entries are always prepared with great care in accordance with the show schedule and, most importantly, the show rules.

So, you see, there is a lot more to bees and beekeeping than you thought! The best showman is one who can take two jars of honey at random from his stock and win a first prize.

Remember this

If you wish, your interest in bees can spread widely into associated activities. If you are competitive, why not enter your local honey show?

Try it now

Numerous activities are related to beekeeping, so you can widen your interest beyond looking after the bees themselves.

Honey, beeswax and propolis can be regarded as by-products of the colony. Each can be used in a variety of different ways, so expand your interest by making things from different hive products.

Focus points

* Hive products, such as honey, beeswax and propolis, can be used in a variety of ways from baking to making polish and candles.
* Beeswax is a valuable hive product that is worth collecting and storing until you have sufficient for your project.
* Beeswax candles can be made simply from a sheet of foundation.
* Skep makers offer courses for you to make your own for collecting swarms.
* There are many bee-related items that you can collect.
* Other associated activities include photography, travel and honey shows.

Next step

The next chapter covers the extremely important topic of health and safety. This is of paramount importance, particularly if you are keeping bees on a rooftop. All beekeepers need to be aware of the risk of anaphylaxis, how to recognize it and what measures to take. The importance of swarm control and keeping good-tempered bees is stressed for the urban situation.

Health and safety

In many ways, this chapter could be regarded as one of the most important in this book. I am not a health and safety obsessive but it is simply common sense to make sure that your beekeeping and your bees are as safe as possible. Let's face it, feeling unsteady when you are carrying a heavy box of honey isn't good for the nerves. Some of these points have been made already, but in this chapter I have attempted to draw the more important ones together.

Anaphylaxis awareness

Anaphylaxis is the extreme allergic reaction to a particular substance; for a beekeeper it is a bee sting or possibly propolis. However well you may be protected and however careful you are, there is always a chance that a bee may sneakily find its way into your veil or sting you through your bee suit.

Anaphylactic shock affects breathing and circulation and someone reacting badly to a bee sting can collapse and, in extreme cases, die. This can happen rapidly and without warning, even if the affected person has never encountered problems previously. While this is a very rare occurrence, you need to be aware of the possibility and know what to do in this situation. (A heart attack or a stroke is a more likely event, so make sure that you know the symptoms for these as well and the action you need to take.)

Working in the apiary alone puts you at a higher risk. If you are going to the apiary on your own, tell someone where you are going and how long you expect to be. Don't forget to let them know when you get back or you could start an unnecessary panic. Take a mobile phone with you. Make sure that the batteries are well charged and put '999' on a speed-dial key. Keep it with you in the apiary. It is no use if you fall over by a hive and your phone is in the car.

You may not be able to get the phone out of your pocket or dial a number on a touch-screen phone while wearing gloves. The emergency services may not be able to come into the apiary for their safety, so having a buddy with you is a very good idea. The emergency services may also be hampered if the access to your apiary is difficult.

Questions you need to ask are:

▶ Could I perform resuscitation through a veil?

▶ Is there an adrenalin pen immediately available and, if so, do I know how to use it? Is it up to date?

Some local associations run courses on dealing with anaphylaxis in the apiary. If yours is among them, I strongly recommend you attend. It will be time and money well spent, even if you never have to deal with the situation. Taking antihistamine an hour before going to the apiary may help to alleviate the reaction to a sting.

If you react increasingly badly to stings, talk to your doctor. At the very least you may need to carry an antihistamine injection. At worst, you may be advised to stop keeping bees. Only you can decide but, personally, I would listen to my doctor. There are desensitization treatments if you really don't want to give up.

Key idea

If you go to the apiary alone, tell someone your plans. Always carry a charged mobile phone with you in case of emergencies; don't leave it in your vehicle.

Remember this

If you develop signs of increasing allergic reaction to bee stings, consult your doctor urgently.

Avoiding slips, trips and falls

When in the apiary, take care not to trip or fall. Don't leave bits of equipment lying around. Not only is this untidy and hazardous, but if you keep bees on a roof, they could be blown off by the wind. Make sure that you wear footwear that has a good grip. Take things slowly and carefully, especially on uneven ground. On a roof, take particular care if there are raised ridges across it.

Remember this

Be careful where you put down your smoker and other equipment.

LIFTING

A full super of honey is heavy. So is a brood box full of bees. Bend your knees and lift with your back rather than bending over and relying on your arms alone. Hive boxes are awkward to carry and it is easy to lose your balance and fall, especially with a design such as the Langstroth which has only small notches as hand (or finger) holds. This is where the National hive comes into its own, being much easier to pick up and carry.

ROOFTOP SAFETY

Moving boxes and equipment up to a roof and back down can be especially hazardous, so always get help for this operation. If the access is narrow or up a ladder, you should never attempt heavy manoeuvres on your own. If you are removing the honey crop, you may have to use a smaller box, such as a nucleus box, to transport a few frames at a time.

Plan your operation carefully. If your apiary is on the roof of a third-party property, such as a hotel or office block, work out the route you are going to take. Try to arrange to transport the supers through the building when few people are there. Saturday would be a good time for an office-block apiary. Almost inevitably, some bees are left in the super or clinging to your bee suit, so you need to avoid areas where people are working. If this isn't possible, warn them what is happening so that those who may be allergic, or just don't want to be around, can keep out of the way. Cover the super and strap it tightly to minimize the chance of escapees.

Make sure that boxes or other heavy equipment are secured tightly when you are carrying them up or down. The last thing you want is for something to slip or come apart, especially if the box is full of bees.

Key idea

When choosing an apiary site, especially on a rooftop, pay particular consideration to access.

Remember this

If your roof is covered with asphalt or a tar-type covering, be careful; it could pose a hazard, especially when you are carrying something heavy.

Keeping cool

On a summer's day, beekeeping can be a very hot occupation. Working in a bee suit can be uncomfortable and cause you to overheat. Just as your bees need water, so do you if you are to avoid becoming dehydrated. This is particularly relevant for rooftop apiaries where the temperature can be significantly higher than at ground level. The roof can also absorb heat, adding to the problem. If you can paint it white, that will help to reflect the heat away. As well as always keeping drinking water on hand, I find using a head sweatband makes life more comfortable and other beekeepers I know also use wrist sweatbands.

Positioning hives safely

In any apiary, it is important to position hives carefully. At ground level, take particular note of neighbours or public-access paths nearby. Neighbours are not such a problem at roof level, but hives here need to be positioned away from the edge. Place them near load-bearing walls and over roof joists. Before you decide to keep bees on a roof, first make sure that it is sound and strong enough to take the weight of a hive containing a full set of frames, a large colony of bees and several supers full of honey. You don't want to give those in the room below an unwelcome surprise!

Make sure that your hive stands are strong enough to take this sort of weight. On soft ground, placing them on a large slab spreads the weight. On a roof, you also need to spread the weight over the roof joists. Strengthen the areas that you will be using the most with some sort of padding, again spreading the weight. A skid-proof covering is also helpful.

If you decide to have visitors to your apiary, limit numbers so that you have sufficient space for them to actually see what is happening. Be sure to provide each with a veil. On a roof, consider the weight that it can bear.

Key idea

It is your responsibility to ensure not only your own safety but also that of your neighbours and passers-by, so you must make sure that your bees are not a nuisance.

CONSIDERING THE WIND

Wind is always a factor in apiary sites. On a roof, you will encounter the additional problem of gusting because the neighbouring properties can form wind corridors. Strap hives down firmly using weatherproof straps made from a material that does not degrade in sunlight. Just putting a couple of bricks on top may work for a large part of the time but think ahead to that unexpected gale. It's better to be safe than sorry and you most certainly don't want anything to be knocked or blown off the roof. Check that your insurance covers you for any type of accident you can imagine.

CONTROLLING SWARMING

As has already been said, swarming needs to be controlled in urban situations as much as is humanly possible. This really is a vital part of your beekeeping and one that you should spend time on until you understand why it happens, how it happens and what steps you can take to prevent and control it. Swarming may be natural bee behaviour but I really do not believe that bees can be left to their own devices in an urban situation.

If you are called to deal with a swarm, don't take risks, particularly in a rooftop situation. No swarm of bees is worth a broken leg, or worse.

Neighbours may not be such a problem if your hives are on a roof but if this is someone else's roof, be aware that maintenance men may need to go up there periodically. Make friends with them, too, as they could be the first to warn you of a problem.

CONSIDERING BEE TEMPERAMENT

Keeping bad-tempered bees simply isn't worth it. You really do need to keep good-tempered bees in any place where other people can come close to your hives. If your bees become bad tempered for any reason, talk to an experienced beekeeper about requeening them, preferably with a locally bred queen and definitely with one that is good tempered.

Consider erecting a sign warning non-beekeeping visitors that there are bees around. Don't fall out with your neighbours or your landlord because they are bothered by your bees. Life is too short for such disputes.

Enjoying your beekeeping

Beekeeping is meant to be an enjoyable hobby. Believe me, with the right bees in the right location, it is. Do yourself a favour and plan out your beekeeping carefully before you start. Learn as much as you can and take advantage of opportunities to visit beekeepers and look at colonies before you get your own. You will make mistakes. We all do – and still do after many years' experience. However, if you learn from those mistakes and rectify them, your experience can only get more pleasurable.

There can be nothing nicer than sitting in the apiary on a summer's evening watching foragers returning, heavily laden, to the hive and hearing the contented hum as the house bees work at turning the nectar into honey.

Sit there with a glass of mead – home-produced, of course – and you will be in paradise!

 Try it now

* Take all necessary measures to remain as safe as possible in the apiary.
* Position hives carefully, being aware of their proximity to neighbours and passers-by.
* If you position hives on a roof, make sure that it is strong enough to support the maximum weight anticipated.
* Protect hives from the wind, particularly on a rooftop where there may be gusting because of surrounding buildings.
* Understand and learn your method of swarm control and then practise it.
* Keep good-tempered bees.

Focus points

* Health and safety is very important, especially when working with livestock (bees), which can be unpredictable.
* Choose apiary sites carefully after considering their position and your access, especially when carrying heavy boxes that may contain bees.

Next step

This book has provided only an introduction to bees and beekeeping. The next section gives references to many other sources where you can learn more.

Further information

This section lists other useful books and magazines on the subject of bees and beekeeping, tells you where to go for bee disease insurance, and gives details of beekeeping associations, beekeeping equipment suppliers and government agencies.

BOOKS ON BEES AND BEEKEEPING

There are many useful books on bees, beekeeping and associated subjects. I have listed a few in each area but, for a wider selection, contact the bee booksellers and appliance manufacturers. Some of the books listed are out of print but it is worth trying to obtain them. Your local beekeeping association may well have copies in its library or you may be able to borrow them from your public library.

Andrews, S. W. (1982) *All About Mead*, Mytholmroyd, Northern Bee Books.

Aston, D. and Bucknall, S. (2004) *Plants and Honey Bees: Their Relationships*, Mytholmroyd, Northern Bee Books.

Badger, M.(2009) *How to Use a Horsley Board for Swarm Control*, Stoneleigh, Bee Craft Ltd.

Bailey, L. and Ball, B. V. (1991) *Honey Bee Pathology*, 2nd edn, London, Academic Press.

Bee Craft Ltd (2005) *The Bee Craft Apiary Guides to Bee Diseases*, Stoneleigh, Bee Craft Ltd.

Bee Craft Ltd (2010) *The Bee Craft Apiary Guide to Colony Make-up: Brood and Adults*, Stoneleigh, Bee Craft Ltd.

Bee Craft Ltd (2005) *The Bee Craft Apiary Guides to Integrated Pest Management*, Stoneleigh, Bee Craft Ltd.

Bee Craft Ltd (2007) *The Bee Craft Apiary Guides to Swarming and Swarm Control*, Stoneleigh, Bee Craft Ltd.

Brown, R. H. (1989) *Beeswax*, 2nd edn, Burrowbridge, Bee Books New and Old.

Canadian Association of Professional Apiculturalists (2000) *Honey Bee Pests and Diseases*, 2nd edn, Guelph, Canadian Association of Professional Apiculturalists.

Caron, D. M. and Connor, L. J. (2013) *Honey Bee Biology and Beekeeping*, rev. edn, Kalamazoo, Wicwas Press.

Chapman, N. (2015) *Pollen Microscopy*, Halstead, Essex, CMI Publishing Ltd.

Chapman, R. (2009) *How to Make a Warming Cabinet for Two Honey Buckets*, Stoneleigh, Bee Craft Ltd.

Chapman, R. (2009) *How to Make an Open-mesh Floor*, Stoneleigh, Bee Craft Ltd.

Collinson, Clarence H. (2003) *What Do You Know?*, Medina, OH, A. I. Root Co.

Cowley, M. (2016) *The Honey Bee Illustrated*, Stoneleigh, Bee Craft Ltd.

Cramp, D. (2008) *A Practical Manual of Beekeeping*, Spring Hill, How To Books.

Davis, C. F. (2004) *The Honey Bee Inside Out*, Stoneleigh, Bee Craft Ltd.

Davis, C. F. (2007) *The Honey Bee Around and About*, Stoneleigh, Bee Craft Ltd.

Ellis, H. *Spoonfuls of Honey*, London, Pavillion Books.

Falk, S. (2015) *Field Guide to the Bees of Great Britain and Ireland*, London, Bloomsbury.

Free, J. B. (1987) *Pheromones of Social Bees*, London, Chapman & Hall.

Flottum, K. (2005) *The Complete and Easy Guide to Beekeeping*, London, Apple Press.

Flottum, K. (2009) *The Honey Handbook*, London, Apple Press.

Gould, J. L. and Gould, C. G. (1988) *The Honey Bee*, New York, Scientific American Library.

Hansen, H. (n.d.) *Honey Bee Brood Diseases*, Copenhagen, L. Launs.

Hodges, D. (1984) *The Pollen Loads of the Honeybee: A Guide to Their Identification by Colour and Form*, Cardiff, International Bee Research Association.

Hooper, T. (1998) *Guide to Bees and Honey*, 3rd rev. edn, Totnes, Marston House Publishers.

Jones, S. L. and Martin, S. J. (eds) (2007) *Apicultural Research on Varroa*, Cardiff, International Bee Research Association.

Kirk, W. (2006) *A Colour Guide to the Pollen Loads of the Honey Bee*, 2nd edn, Cardiff, International Bee Research Association.

Lindauer, M. (1978) *Communication among Social Bees* (Harvard Books in Biology, No. 2), 2nd edn, Cambridge, MA, Harvard University Press.

Morse, R. A. and Flottum, K. (eds) (1997) *Honey Bee Pests, Predators and Diseases*, 3rd edn, Medina, OH, A. I. Root Co.

Munn, P. (ed.) (1998) *Beeswax and Propolis for Pleasure and Profit*, Cardiff, International Bee Research Association.

Riches, H. (2003) *Insect Bites and Stings: A Guide to Prevention and Treatment*, Cardiff, International Bee Research Association.

Riches, H. (1997) *Mead: Making, Exhibiting and Judging*, Burrowbridge, Bee Books New and Old.

Sammataro, D. (2014) *Diagnosing Bee Mites with Emphasis on Varroa*, Mytholmroyd, Northern Bee Books.

Sammataro, D. and Yoder, J. A. (eds) (2012) *Honey Bee Colony Health: challenges and sustainable solutions*, New York, CRC Press.

Schaker, M. (2008) *A Spring without Bees*, Guilford, CT, The Globe Pequot Press.

Schramm, K. (2003) *The Compleat Meadmaker*, Boulder, CO, US, Brewers Publications.

Seeley, T. D. (2010) *Honeybee Democracy*, Princeton, NJ, Princeton University Press.

Seeley, T. D. (1995) *The Wisdom of the Hive: The Social Physiology of Honey Bee Colonies*, Cambridge, MA, Harvard University Press.

Tautz, J. (2008) *The Buzz about Bees*, Berlin, Springer-Verlag.

Vidal-Naquet, N. (2015) *Honeybee Veterinary Medicine: Apis mellifera L*, Sheffield, 5m Publishing.

Von Frisch, K. (1967) *The Dance Language and Orientation of Bees*, Cambridge, MA, Harvard University Press.

Waring, A. (2004) *Better Beginnings for Beekeepers*, 2nd edn, Doncaster, BIBBA.

Wedmore, E. B. (1979) *A Manual of Beekeeping*, 2nd rev. edn, Burrowbridge, Bee Books New and Old.

White, J. and Rogers, V. (2000) *Honey in the Kitchen*, rev. edn, Charlestown, Bee Books New and Old.

White, J. and Rogers, V. (2001) *More Honey in the Kitchen*, rev. edn, Charlestown, Bee Books New and Old.

Wilson, J. S. and Messinger Carril, O. (2015) *The BEES in your Backyard: a guide to North America's bees*, Princeton and Oxford, Princeton University Press.

Wilson-Rich, N. (2014) *The Bee: A natural history*, Lewes, Ivy Press.

Winston, M. L. (1987) *The Biology of the Honey Bee*, Cambridge, MA, Harvard University Press.

BEE BOOK SUPPLIERS

C. Arden Bookseller, Radnor House, Church Street, Hay-on-Wye, Hereford HR3 5DQ (www.ardenbooks.co.uk)

International Bee Research Association, Hendal House, Hendal Hill, Groombridge, East Sussex CF37 5YR (www.ibra.org.uk)

Northern Bee Books, Scout Bottom Farm, Mytholmroyd, Hebden Bridge, West Yorkshire HX7 5JS (www.northernbeebooks.co.uk)

BEE DISEASE INSURANCE

Colonies can be insured against losses resulting from European foul brood and American foul brood with Bee Diseases Insurance Ltd, c/o British Beekeepers' Association, National Beekeeping Centre, Stoneleigh Park, Kenilworth, Warwickshire CV8 2LG (www.bbka.org.uk).

BEEKEEPING ASSOCIATIONS

England: The British Beekeepers' Association, National Beekeeping Centre, Stoneleigh Park, Kenilworth, Warwickshire CV8 2LG (www.bbka.org.uk)

Scotland: The Scottish Beekeepers' Association (www.scottishbeekeepers.org.uk)

Northern Ireland: The Ulster Beekeepers' Association (www.ubka.org)

The Institute of Northern Ireland Beekeepers (www.inibeekeepers.com)

Republic of Ireland: The Federation of Irish Beekeepers' Associations (www.irishbeekeeping.ie)

Wales: The Welsh Beekeepers' Association (www.wbka.com)

BEEKEEPING EQUIPMENT SUPPLIERS

Agri-Nova Technology Ltd, The Old Forge, Wendens Ambo, Saffron Walden, Essex CB11 4JL (www.agri-nova.biz)

Andermatt BioVet AG, Stahlermatten 6, CH-6146 Grossdietwil, Switzerland (www.biovet.ch)

BB Wear, 1 Glyn Way, Threemilestone, Truro, Cornwall TR3 6DT (www.bbwear.co.uk)

Bee Basic, 5 Hillcrest Avenue, Pinner, Middlesex HA5 1AJ (www.beebasic.co.uk)

Bee Equipped, Brunswood Farm, Brunswood Lane, Bradley, Ashbourne, Derbyshire DE6 1PN (www.beeequipped.co.uk)

Bee Proof Suits, Bowns Hill, Matlock, Derbyshire DE4 5DG (www.beeproofsuits.com)

BJ Sherriff International, South Cornwall Honey Farm, Carclew, Mylor, Falmouth, Cornwall TR11 5UN (www.bjsherriff.com)

Blue Bell Hill Apiaries, Ivy Farm, Lidsing Road, Lidsing, Nr Gillingham, Kent ME7 3NL (www.bbha.biz)

Brunel Microscopes Ltd, Unit 2, Vincients Road, Bumper's Industrial Estate, Chippenham SN14 6NQ (www.brunelmicroscopes.co.uk)

C. Wynne Jones, Ty Brith, Pentre Celyn, Ruthin LL15 2SR (www.beesupplies.co.uk or www.bottlesandjars.co.uk)

Caddon Hives, Caddon Lands, Mill Bank Road, Clovenfords, Galashiels TD1 3LZ (www.caddon-hives.co.uk)

Chemicals Laif S. p. A., Industria Bio-chimica, Via dell'artigianato, 13, 35010 Vigonza (PD), Italy.

Circomb, 29 Glamis Road, Dundee DD2 1TS

Compak (South) Ltd, 3 Ashmead Road, Keynsham, Bristol BS31 1SX (www.compaksouth.com)

E. H. Thorne (Beehives) Ltd, Beehive Works, Wragby, Market Rasen LN8 5LA (www.thorne.co.uk)

Freeman and Harding Ltd, Unit 18, Bilton Road, Erith DA8 2AN (www.freemanharding.co.uk)

Harrison Smith French Flint Ltd, Unit 4G, The Leathermarket, London SE1 3ER (www.frenchflint.com)

Kemble Bee Supplies, Brede Valley Bee Farm, Cottage Lane, Westfield, Hastings TN35 4RT (www.kemble-bees.com)

Maisemore Apiaries, Old Road, Maisemore, Gloucester GL2 8HT (www.bees-online.co.uk)

Mann Lake UK, The Pack House, Highland Court
Farm, Coldharbour Lane, Bridge, Canterbury CT4 5HN
(www.mannlake.co.uk)

Modern Beekeeping, Rooftops, Ebrington Street, Kingsbridge
TQ7 1DE (www.modernbeekeeping.co.uk)

National Bee Supplies, Merrivale Road, Exeter Road
Industrial Estate, Okehampton, Devon EX20 1UD
(www.beekeeping.co.uk)

Park Beekeeping Supplies, Unit 17 Blackheath Business
Estate, 78B Blackheath Hill, London SE10 8BA
(www.parkbeekeeping.com)

Paynes Southdown Bee Farms, Bentley Cottage, Wickham Hill,
Hassocks, West Sussex BN6 9NP (www.paynesbeefarm.co.uk)

Stamfordham Ltd, Heugh House, Heugh, Newcastle upon Tyne
NE18 0NH (www.stamfordham.biz)

Vita (Europe) Ltd, Vita House, London Street, Basingstoke
RG21 7PG (www.vita-europe.com)

BEEKEEPING MAGAZINES
An Beachaire, published by the Federation of Irish Beekeepers'
Associations (www.irishbeekeeping.ie)

BBKA News, published monthly by the British Beekeepers'
Association (www.bbka.org.uk; available only as part of
Association Membership)

Bee Craft, published monthly by Bee Craft Ltd
(www.bee-craft.com)

Beekeepers Quarterly, published quarterly by Northern Bee
Books (http://beekeepers.peacockmagazines.com)

Gwenynwyr Cymru, published quarterly by the Welsh Beekeepers' Association (www.wbka.com)

The Scottish Beekeeper, published by the Scottish Beekeepers' Association (www.scottishbeekeepers.org.uk; available only as part of Association membership)

Most local beekeeping associations also publish their own magazine or newsletter.

GOVERNMENT AGENCIES

England and Wales: The National Bee Unit, The Animal and Plant Health Agency, Sand Hutton, York YO41 1LZ (01904 462510; beebase.csl.gov.uk)

Scotland: Director General Environment, The Scottish Government (0131 556 8400)

Northern Ireland: The Department of Agriculture and Rural Development Northern Ireland (028 9052 4420)

Southern Ireland: Teagasc, Bee Diagnostic Unit, Malahide Road, Dublin 17, Ireland

UK: Food Standards Agency, website: hzttp://www.food.gov.uk/foodindustry/regulation/foodlawguidebranch/foodlawguidech03/foodlawhoney.

USA: US Food and Drug Administration (www.fda.gov)

Glossary

Abdomen The third and largest body segment of the bee, which contains the heart, stomach and intestines. In the worker, it contains the sting and wax glands. In the drone it contains the testes. In the queen it contains the ovaries and the spermatheca.

Acarine A disease caused by a mite (*Acarapis woodi*), which infests the bee's tracheae leading from the first pair of spiracles on the thorax. Known as the Isle of Wight disease because if was first noted there in 1906.

Alighting board A strip of wood, usually fixed to the hive stand or as part of the floor, which protrudes in front of the entrance, giving a platform on which bees can land before running into the hive.

American foul brood (AFB) A disease caused by a spore-forming bacterium (*Paenibacillus larvae* subspecies *larvae*) that develops in the gut of the larva and kills it after the cell is sealed. This causes sunken and perforated cappings. It is a notifiable disease.

Anaphylactic shock A severe reaction resulting from an acute allergy to bee venom. It may cause sudden death unless immediate medical attention is received.

Antenna (plural: antennae) One of a pair of 'feelers' on the head of the bee which carries sensory cells for touch, smell and vibration.

Api-Bioxal® An oxalic acid treatment for control of *Varroa destructor*.

Apiary The place where one or more hives are kept.

Apiculture The practice of keeping bees.

Apiguard® A thymol-based treatment for control of *Varroa destructor*.

ApiLife Var® A treatment for controlling *Varroa destructor* based on thymol, menthol, eucalyptus and camphor.

Apistan® A slow-release polymer strip pyrethroid formulation specifically designed to control *Varroa destructor*.

Ashforth feeder A wooden feeder the same size as the hive box. There is one syrup reservoir to which bees gain access from one side.

Asian hornet See '*Vespa velutina*'.

Bait hive A hive placed to attract stray swarms.

Bayvarol® A slow-release polymer strip pyrethroid formulation specifically designed to control *Varroa destructor*.

Beehive A container for housing honeybees consisting of a floor, brood box, one or more supers, an inner cover and a roof.

Bee space The space left by bees between the comb and other surfaces in the hive. It is large enough for queen, worker and drone to pass through.

Bee suit A suit worn by a beekeeper for protection and comfort while opening beehives or collecting swarms.

Bee-tight The situation where there is no way into the hive other than the entrance.

BeeVital Hive Clean® A treatment to help colonies remain strong and therefore more able to overcome the effects of *Varroa destructor*.

Beeswax A hydrocarbon produced from glands on the underside of the abdomen of the worker bee. Used by bees for comb building and capping cells.

Brace comb Bridges of wax built between adjacent surfaces in the hive.

Brood The immature life stages of the bee. Cells containing eggs and larvae are known as open brood. Capped cells in which the larvae pupate into adult bees are known as sealed brood.

Brood box The area in which the queen is confined and the brood is reared.

Brood pattern The pattern of concentric swirls of brood at different stages of development. A good brood pattern has few empty cells and indicates that the queen's brood is largely healthy.

Burr comb Wax built on a comb or on a wooden part in a hive but not connected to any other part.

Canadian escape A board with no moving parts for clearing bees from honey supers. Bees pass through narrow gaps and fail to find a way back.

Capped brood See 'Sealed brood'.

Cappings The beeswax covering over a cell. Cappings over honey consist of wax only. Those over brood include hair and other materials.

Cast The same as a swarm in all respects except that it may contain one or more unmated queens.

Caste A different form of the same sex. Bees have two castes in the female form – queen and worker.

Castellations Thin pieces of metal into which slots are cut to take frame lugs. Fastened to the inside upper edge of the hive body from which the frames are suspended. Designed to maintain a constant spacing between frames. Available with differing numbers of slots.

Cells Small, six-sided hexagonal wax compartments making up honeycomb. Used to store honey and pollen and to rear the juvenile life stages of bees.

Chalk brood A disease caused by a fungus (*Ascosphaera apis*) that affects sealed brood.

Clearer board An inner cover designed to accommodate one or two Porter bee escapes (q.v.).

Cleansing flight One made by bees that have been confined to the hive for long periods such as in winter or during bad weather. Bees avoid defecating inside the hive and make a cleansing flight when the weather improves.

Cold way Frames arranged at right angles to the entrance of a hive. See also 'Warm way'.

Colony The viable living unit for honeybees comprising a queen and workers. During the summer, male drones are also present.

Comb The mass of six-sided beeswax cells built by honeybees in which brood is reared and honey and pollen are stored.

Contact feeder One that gives bees direct contact with the contents. It must be surrounded by an eke or empty super so that the roof can be replaced tightly.

Corbicula (plural: Corbiculae) See 'Pollen basket'.

Crystallization See 'Granulation'.

Cut comb Natural comb or comb built on thin foundation cut to a size to fit a container for sale.

Cut comb foundation Sheets of beeswax foundation as close as practicable to the thickness of the midrib found in naturally built comb.

Dextrose See 'Glucose'.

Drawn comb Foundation where the cells have been drawn out by the worker bees into full-depth cells.

Drifting The tendency of bees from one colony to accidentally enter another when returning from foraging flights.

Drone The male bee whose main function is to fly to drone assembly areas and mate with a virgin queen.

Drone comb Sections of the comb built for raising drones. The cells are slightly larger than worker cells and have a convex, domed capping when sealed.

Drone congregation area The place where drones congregate to mate with virgin queens, which travel to the same area.

Drone layer A queen that lays only unfertilized eggs, which develop into drones.

Dummy frame A slab of wood cut to the same size as a frame to take its place in a hive.

Dysentery A condition caused by an excessive amount of water in a bee's body, usually as a result of prolonged confinement during winter and early spring and consumption of food with a high water content. Often, but not invariably, associated with nosema.

Egg The first stage of honeybee metamorphosis. Eggs laid by the queen appear as small, thin, rods, about 1.6 mm ($^1/_{20}$ in.) long, usually placed in the bottom of the cell.

Eke Four pieces of wood nailed together into a rectangle the same size as the hive. Used to extend the hive when required.

Entrance The elongated space across the front of a beehive through which bees exit and enter the hive.

Entrance block A removable block of wood used to reduce the width of the hive entrance.

European foul brood (EFB) A disease caused by a bacterium (*Melissococcus plutonius*) that infects the gut of the developing larva and competes for food. Does not kill all affected larvae. A notifiable disease not confined to Europe.

Exomite™ Apis A treatment to help colonies remain strong and therefore more able to overcome the effects of *Varroa destructor*.

Exoskeleton The hard outside covering of all insect bodies, including bees.

Fermentation The chemical breakdown of honey, caused by sugar-tolerant yeast and associated with honey having a high moisture content. Used to advantage when making mead.

Fertile queen A mated queen that can lay fertilized eggs.

Flying bees Worker bees old enough to have largely completed their duties in the hive that go out foraging for nectar and pollen. Foraging generally starts at three weeks of age.

Following The habit of some bees to follow and possibly sting another animals coming near their nest.

Foragers See 'Flying bees'.

Foraging The act of seeking for and collecting nectar, pollen, water and propolis.

Foundation Beeswax sheets impressed with the shape of cell bases. Available in sizes suitable for worker and drone cells, it can be strengthened with wires or used without.

Frame Wooden or plastic structure designed to hold bee comb and enable the beekeeper to inspect and utilize it fully.

Frame runner A narrow piece of folded metal fastened to the inside upper edge of the hive body from which the frames are suspended.

Frame spacers Plastic or metal spacers which fit over frame lugs and butt up to the spacer on the adjacent frame to ensure constant spacing. Can be narrow or wide. Wide spacing is used only in supers where deeper honey-storage cells are desired.

Fructose The predominant simple sugar found in honey. Also known as levulose.

Glucose One of the two principal sugars that constitute honey. Also known as dextrose.

Granulation When crystals are formed naturally in honey by the least soluble sugar (glucose), especially when its temperature falls.

Guard bees Bees at the hive entrance that guard it from invaders. Guard bees give off an alarm pheromone if the hive is disturbed or threatened and are the first to fly at and attack the invader.

Hefting The act of lifting a hive slightly from its support to ascertain its weight.

Hive An artificial structure intended as a home for bees. The best hives allow beekeepers to inspect all aspects of bee life.

HiveAlive™ A treatment to help colonies remain strong and therefore more able to overcome the effects of *Varroa destructor*.

Hive stand A structure that supports the hive and raises it off the ground.

Hydroxymethyfurfural (HMF) An organic compound in honey the level of which rises with age and heating. Levels in honey for sale are defined in legislation.

Hive tool The composite lever/scraper used in the manipulation of a colony.

Hoffman frame A type of self-spacing frame.

Honey The concentrated form of nectar that will keep for a long time. Its colour and flavour depend on the flowers from which the nectar is gathered.

Honey crop An organ in the bee's abdomen used for carrying nectar, honey or water.

Honey extractor Machine that allows honey to be extracted from combs so that they can be reused.

Honey flow A heightened influx of nectar into the hive, brought about by favourable weather conditions and the availability of suitable flowers.

Honey ripener See 'Settling tank'.

Honeydew The product of sap-sucking bugs such as aphids. Collected by bees when it is diluted by dew.

House bee A young worker that stays in the hive and performs tasks such as feeding young larvae, cleaning cells, and receiving and storing nectar and pollen from foragers.

Inner cover A board that is placed over the frames just beneath the roof.

Integrated Pest Management (IPM) The use of substances and specific manipulations to reduce the population of *Varroa destructor*.

Invert sugar syrup A liquid sugar syrup formed by inversion, or chemical breakdown, of sucrose resulting in an equal mixture of glucose (dextrose) and fructose (levulose).

IPM See 'Integrated Pest Management'.

June gap A period during June when there is a serious shortage of forage.

Larva The second stage of bee metamorphosis. The larva hatches from the egg. It then develops into a pupa and changes into an adult.

Laying worker A worker that lays unfertilized but fertile eggs, producing only drones. This occurs if a colony becomes queenless and is not able to raise a new queen.

Levulose See 'Fructose'.

Mating flight The flight taken by a virgin queen when she mates in the air with several drones.

Mead An alcoholic drink made from honey and water.

Melomel A type of mead made using fruit juice and honey.

Metal ends See 'Frame spacers'.

Metal runner See 'Frame runner'.

Miller feeder A wooden feeder the same size as the hive box. There are two syrup reservoirs to which bees gain access from a central slot.

Modified National The commonest hive in use in the UK. It is a single-walled hive.

Moult The shedding of skin by a larva to allow for its further growth.

Mouseguard A metal strip or similar containing holes that allow bees in and out of the hive but prevent mice from gaining access.

Nectar The sugary secretion of plants produced to attract insects for the purpose of pollination.

Nectar guides Marks on flowers believed to direct insects to nectar sources. They may be visible to the human eye or may reflect ultraviolet and hence be visible only to bees.

Nosema A disease caused by microsporidian parasites (*Nosema apis* and *Nosema ceranae*). These infect the gut of the bee and shorten its life by preventing it from digesting its food properly.

Nucleus A small hive designed to contain three, four or five frames only.

Nucleus hive A small colony, usually on three, four or five frames. Used primarily for starting new colonies, or for rearing or storing queens. Also known as a 'nuc'.

Nurse bees Young worker bees, three to ten days old, which feed and take care of developing brood.

NutriBee® A treatment to help colonies remain strong and therefore more able to overcome the effects of *Varroa destructor*.

Orientation flight A short flight taken by a young worker in front of or near the hive prior to when she starts foraging, in order to establish the position of the hive.

Out-apiary Apiaries established away from the beekeeper's home.

Pheromone A substance produced by one living thing that affects the behaviour of other members of the same species. Pheromones produced by the queen help the colony to function properly.

Plastic ends See 'Frame spacer'.

Play cells See 'Queen cell cups'.

Play flight See 'Orientation flight'.

Pollen The male reproductive part of the plant.

Pollen basket (corbicula) A segment on the hind pair of legs in a worker bee specifically designed for carrying pollen. Also used to bring propolis back to the hive.

Pollen load The pellets of pollen carried by a foraging worker bee in the pollen baskets (corbiculae) on its hind pair of legs.

Pollination The transfer of pollen from the anthers to the stigma of flowers.

Porter bee escape A device for clearing bees from supers. Two spring valves allow bees to pass through one way but not return.

Prime swarm The first swarm to leave the colony, usually containing the old queen.

Proboscis The mouthparts of the bee that form the sucking tube or tongue. Used for sucking up liquid food (nectar or sugar syrup) or water.

Propolis A resinous material collected by bees from the opening buds of various trees, such as poplars.

Pupa The third stage in the development of the honeybee, during which the organs of the larva are replaced by those that will be used by an adult. Takes place in a sealed cell.

Queen One of the two variants or castes of the female in bees. Larger and longer than the worker bee.

Queen cell cups The base of a queen cell into which the queen will lay an egg designed to develop into a new queen.

Queen cell An elongated brood cell hanging vertically on the face of the comb in which a queen is reared.

Queen excluder A device with slots or spaced wires, which allows workers to pass through but prevents the passage of queens and drones.

Queen substance Complex pheromones produced by the queen. Transmitted throughout the colony through the exchange of food between workers, to alert other workers of the queen's presence. Its presence stops worker bees rearing more queens and/or inhibits them from laying eggs.

Queenless The situation when a colony has no queen. If bees have access to worker eggs or very young larvae, they are able to rear a replacement queen.

Queenright The situation when a colony has a living and laying queen.

Retinue Worker bees that attend the queen and care for her needs within the hive.

Ripe queen cell A queen cell that is near to hatching. Bees remove wax from the tip, exposing the brown parchment-like cocoon.

Robbing When wasps, or bees from other colonies, try to steal honey from a hive.

Royal jelly A highly nutritious glandular secretion of young bees, used to feed the queen, young brood and larvae being reared as new queens.

Sacbrood A virus disease thatprevents the final larval moult. The larva dies in its larval skin, which is easily removed from the cell.

Scout bees Worker bees that search for new sources of nectar, pollen, water and propolis. If a colony is preparing to swarm, scout bees will search for a suitable location for the colony's new home.

Sealed brood The pupal stage in a bee's development during which it changes into an adult.

Sections Honeycomb built into special basswood frames. Generally sold complete. Also available in circular plastic form.

Self-spacing frame A frame in which the upper part of the side bar is extended to touch that of the adjacent frame. Designed to maintain a constant distance between adjacent frames.

Settling tank A holding tank for honey that allows air to rise to the surface before bottling using the tap at the base.

SHB See 'Small hive beetle'.

Shook swarm A mass of bees shaken, together with their queen, from one hive into another. Used to control swarming or diseases such as *Varroa destructor* or European foul brood.

Skep A beehive constructed from straw that does not contain moveable frames. Often used for collecting swarms.

Small hive beetle (SHB) A small beetle (*Aethina tumida*) about one-third the size of a worker bee. Dark red, brown or black, with distinctive clubbed antennae. Both larvae and adults eat honey and pollen. Will spoil honey in the comb. Not yet thought to be present in the UK. It is a notifiable disease in the UK.

Smoke The product of burning suitable materials. The best smoke for working with bees comes from organic materials such as rotten wood, shavings, dried grass, etc.

Smoker Device that delivers smoke into the hive in a precise manner in order to calm the bees and facilitate the beekeeper's access.

Spermatheca A special organ in the queen's abdomen in which she stores sperm received from drones during mating.

Spiracles Apertures found on the sides of the thorax and abdomen which lead to the breathing tubes or tracheae.

Sting The defensive mechanism at the end of the abdomen that produces venom (q.v.) through a barbed stinger, used by worker bees to deter predators. The queen will use her sting to kill rival queens, usually during the swarming process.

Stone brood A fungal disease similar to chalk brood caused by *Ascosphera flavus*.

Stores The weight of honey collected by bees, especially the reserves needed for winter.

Super The box(es) placed on top of the brood chamber to increase the space available to the colony for honey storage.

Sugar syrup A solution of sugar and water used to feed bees.

Swarm A mass of bees containing a queen, not in a hive.

Swarm cell Queen cells, often but not always found on the bottom of the combs before swarming.

Swarm control Methods used by beekeepers to stop a swarm from leaving the hive.

Swarm prevention Methods used by beekeepers to prevent the physical conditions arising that stimulate a colony to prepare to swarm.

Thin foundation A sheet of foundation thinner than that used for brood rearing. Used for the production of cut-comb honey.

Thorax The second and central part of the bee's body, containing the flight muscles and where the legs and wings are attached.

***Tropilaelaps* mites** Parasitic mites that affect both developing brood and adult honey bees. Not currently known to be in the UK. A statutory notifiable pest of honeybees.

Uncapping knife A knife used to remove the cappings from combs of sealed honey prior to extraction.

Uniting The act of combining two or more colonies to form a larger colony.

Varroa destructor A mite that breeds in sealed brood cells, feeding on the larval blood. Transmits viruses that affect the developing bee.

Veil The see-through, bee-proof garment worn by beekeepers to protect against stings.

Venom Poison secreted by special glands attached to the bee's sting.

Venom allergy A condition in which a person, when stung, may experience a variety of symptoms ranging from a mild rash or itchiness to anaphylactic shock. A person who is stung and experiences abnormal symptoms should consult a doctor before working with bees again.

Vespa velutina The Asian hornet now in France, Belgium and Spain. Predates and can decimate honeybee colonies.

Virgin queen A young, unmated queen.

Waggle dance The most common communication dance used by bees to indicate a food source more than 100 m(330 ft) from the hive.

Warm way Frames arranged parallel to the entrance of a hive. See also 'Cold way'.

Wax glands Eight pairs of glands on the underside of the last four visible abdominal segments of the worker bee, which secrete small particles of beeswax.

Wax moth (greater) The greater wax moth (*Galleria mellonella*) is a serious and destructive pest of unprotected honeycomb. Primarily infests stored equipment but will invade weak colonies. The larvae chew into woodwork to make depressions in which to pupate.

Wax moth (lesser) The lesser wax moth (*Achroia grisella*) is similar to the greater wax moth but causes less damage. The head of an adult is yellow.

WBC hive A double-walled hive designed by William Broughton Carr.

Windbreak A barrier to break the force of the wind blowing on to hives in an apiary. The best windbreak is a thick hedge.

Winter cluster The roughly spherical mass formed by a bee colony as a means to survive the winter.

Worker The commonest bee in the colony. Specialized to undertake the tasks required for the continuation of the colony such as feeding young larvae and foraging for nectar and pollen.

Worker comb Sections of the comb built for raising worker bees. When sealed, the cappings are flat. Also used for storing honey and pollen.

Index

Picture acknowledgements

Figures 2.1, 9.1 and 17.1: © Claire and Adrian Waring

Figure 3.1: © Valery Rizzo / Alamy Stock Photo

Figure 6.1: © Skim New Media Limited / Alamy Stock Photo

Figures 17.2, 18.1, 19.1, 19.2 and 19.3: © The Animal and Plant Health Agency (APHA)